Migration Borders Freedom

International borders have become deadly barriers of a proportion rivaled only by war or natural disaster. Yet despite the damage created by borders, most people can't – or don't want to – imagine a world without them. What alternatives do we have to prevent the deadly results of contemporary borders?

In today's world, national citizenship determines a person's ability to migrate across borders. *Migration Borders Freedom* questions that premise. Recognizing the magnitude of deaths occurring at contemporary borders worldwide, the book problematizes the concept of the border and develops arguments for open borders and a world without borders. It explores alternative possibilities, ranging from the practical to the utopian, that link migration with ideas of community, citizenship, and belonging. The author calls into question the conventional political imagination that assumes migration and citizenship to be responsibilities of nation states, rather than cities. While the book draws on the theoretical work of thinkers such as Ernst Bloch, David Harvey, and Henri Lefebvre, it also presents international empirical examples of policies and practices on migration and claims of belonging. In this way, the book equips the reader with the practical and conceptual tools for political action, activist practice, and scholarly engagement to achieve greater justice for people who are on the move.

Harald Bauder is Professor in the Department of Geography and Environmental Studies, and the Graduate Program for Immigration and Settlement Studies (ISS) at Ryerson University in Toronto, Canada, and the founding Academic Director of the Ryerson Centre for Immigration and Settlement (RCIS).

Routledge Studies in Human Geography

This series provides a forum for innovative, vibrant, and critical debate within Human Geography. Titles will reflect the wealth of research which is taking place in this diverse and ever-expanding field. Contributions will be drawn from the main sub-disciplines and from innovative areas of work which have no particular sub-disciplinary allegiances.

For a full list of titles in this series, please visit www.routledge.com/series/SE0514

56 Ageing Resource Communities
New frontiers of rural population change, community development and voluntarism
Edited by Mark Skinner and Neil Hanlon

57 Access, Property and American Urban Space
M. Gordon Brown

58 The Production and Consumption of Music in the Digital Age
Edited by Brian J. Hracs, Michael Seman and Tarek E. Virani

59 Geographies of Entrepreneurship
Edited by Elizabeth A. Mack and Haifeng Qian

60 Everyday Globalization
A spatial semiotics of immigrant neighborhoods in Brooklyn and Paris
Timothy Shortell

61 Releasing the Commons
Rethinking the futures of the commons
Edited by Ash Amin and Philip Howell

62 The Geography of Names
Indigenous to post-foundational
Gwilym Lucas Eades

63 Migration Borders Freedom
Harald Bauder

64 Communications/Media/Geographies
Paul C. Adams, Julie Cupples, Kevin Glynn, André Jansson and Shaun Moores

65 Public Urban Space, Gender and Segregation
Reza Arjmand

Migration Borders Freedom

Harald Bauder

LONDON AND NEW YORK

First published in paperback 2018

First published 2017
by Routledge
2 Park Square, Milton Park, Abingdon, Oxon OX14 4RN

and by Routledge
711 Third Avenue, New York, NY 10017

Routledge is an imprint of the Taylor & Francis Group, an informa business

British Library Cataloguing in Publication Data
A catalogue record for this book is available from the British Library

Library of Congress Cataloging in Publication Data
Names: Bauder, Harald, 1969- author.
Title: Migration, borders, freedom / Harald Bauder.
Description: Abingdon, Oxon ; New York, NY : Routledge is an imprint of the Taylor & Francis Group, an Informa Business, [2017] | Series: Routledge studies in human geography
Identifiers: LCCN 2016011104| ISBN 9781138195608 (hardback) | ISBN 9781315638300 (e-book)
Subjects: LCSH: Boundaries–Political aspects. | Boundaries–Social aspects. | Border crossing. | Border security.
Classification: LCC JC323 .B38 2017 | DDC 320.1/2–dc23
LC record available at https://lccn.loc.gov/2016011104

ISBN: 978-1-138-19560-8 (hbk)
ISBN: 978-1-138-54499-4 (pbk)
ISBN: 978-1-315-63830-0 (ebk)

Typeset in Times New Roman
by Taylor & Francis Books

For everyone

Contents

Figures

Preface

The title *Migration Borders Freedom* is a play on words. It can be interpreted as three nouns, in which case it describes the key concepts addressed in the book. The publisher was keen to have the book's key concepts included in the main title to "signify the market." Thus, a reader browsing the internet or bookshelf immediately knows that this book is about migration, borders, and freedom.

The book's title can also be interpreted as a sentence, in which case the word "borders" becomes a verb. There are at least two ways in which the title can be read as a sentence: first, it can mean that migration creates a border around freedom. This meaning signifies that migration does not lead to greater freedom but is rather a limitation thereof. It may apply, for example, to people who possessed the freedoms associated with citizenship, economic security, belonging to a community, and protection by the state in their countries of origin, but who lost these freedoms when they migrated to a different country where they lack citizenship, are denied access to the labor market, experience discrimination, and may even be treated as criminals.

The second way of reading the sentence "migration borders freedom" is as "migration is on the border of freedom." I visualize this meaning as a mediaeval European city that is bordered by a fortified wall. Inside the wall reside free citizens; in the surrounding hinterland live serfs who are bonded to their feudal lord. Migration can take the serfs to the gates of the city, which is as close as migration can bring them to freedom. However, it is up to the city's gatekeeper to permit the migrants entry and thus to gain freedom. According to this meaning, migration is *just* outside of the realm of freedom. Migration isn't freedom – it's close to it but not quite there.

The various ways of reading the title reflect the content of the book. On the one hand, the book addresses the problem of how borders and migration are often associated with the denial of rights and freedoms. On the other hand, it acknowledges the prospect that migration offers to gain freedom in the form of rights, protection, belonging, and economic security. The book also searches for solutions that enable migrants to leap over the metaphorical city wall that stands between them and freedom. In fact, the city will reappear as an important figure in the second part of the book, in which I discuss such solutions.

In addition, the ambiguity of the title – that it can be interpreted in various ways – mirrors the approach I chose in this book to explore the concepts of borders, migration, and freedom. These concepts, too, can be interpreted in various ways, depending on the vantage point and interests of the observer. Engaging the various interpretations – and their contradictions – are an integral part of the way in which this book searches for alternative border and migration practices.

Throughout the book, I use the terms migration and migrant. Some of my colleagues are critical of these terms and would argue that "mobility" is a better term to capture the complex patterns of the movement of people across the surface of the earth. Others may suggest that the term "migrant" represents oppressive state practices that created this category and imposed it on human beings in the first place. I decided nevertheless to use the term migrant because, to me, it represents a person who is not only mobile but also lays claims to rights and belonging. The people who Europeans enslaved in Africa from the 15th to the 19th centuries, whose rights were revoked, and who were shipped in chains to the Americas were considered a mobile commodity by the slave traders. These slaves were denied being migrants, which would have granted them their humanity, rights, and free will. By using the term migrant, I acknowledge the humanity, the rights to belong, and free will that people on the move possess.

Migration Borders Freedom is an exploration of ideas. It not only critiques the suffering that border practices are inflicting on migrants; it also seeks practical as well as far-sighted solutions. The book is thus intended to equip its reader with the practical and conceptual tools for political action, activist practice, and scholarly engagement towards greater freedom and justice for migrants.

As I embarked upon writing *Migration Borders Freedom*, I envisaged an audience of academics, researchers, advanced students, activists, and policy makers with an interest in migration and cross-border mobility and who do not shy away from "big" thinking that challenges taken-for-granted ideas. Although my discipline is geography and this book is part of the Routledge Studies in Human Geography series, *Migration Borders Freedom* is an interdisciplinary book that can be read across the spectrum of social sciences and humanities, including anthropology, geography, history, philosophy, political science, and sociology, and in transdisciplinary fields such as border, international, migration, and refugee studies.

To attract a general audience with interests in progressive border and migration politics, I made every effort to write in a jargon-meager manner accessible to non-expert audiences. For most academics, such a writing style does not come intuitively. Often, I think, we shortcut our thought processes by relying on terminology that conveys previously developed ideas that other expert scholars are supposed to be able to decipher. I found that by dropping these academic terms, I could no longer hide behind a veil of jargon that is habitually vague and confusing. Moreover, explaining myself in clearer language forced me to sharpen my stream of thought, which not only benefits the reader, but helped me to refine the logic of my argument. Nevertheless, there were instances when I found it challenging to drop academic jargon. For example, I could not find an effective way to circumvent the term "dialectics," a word that my spouse – an editor by profession – finds eminently repulsive. Instead of avoiding the term, I tried to make sure that the reader understands the meaning in which I apply it. One of my proudest achievements is that I managed to write a book without footnotes or endnotes, which I hope has improved the flow of the ideas within the text.

With examples drawn from different historical periods and geographic locations, *Migration Borders Freedom* is intended for a global readership. The reviewers of the book's prospectus suggested that I include even more historical material to increase the "shelf life" of the book and illustrate the "timeless" nature of the topic. While I agree that the topic of mobility across borders is a perennial issue, the practices and policies involving borders, migration, and freedom are particularly problematic today. These policies and practices have killed and disenfranchised record numbers of people in recent years. While I heeded the reviewers' advice and included some historical material, I wrote the book primarily from a contemporary vantage point. At the time of writing, the topics of migration, border controls, and the infringement of the freedom of mobility captivated the attention of the news and politics. In fact, cross-border migration was *the* dominant issue in the news and the most hotly debated political topic in Germany between mid-2015 and early 2016, when I completed the manuscript.

The reviewers also encouraged me to illustrate my argument by including more examples from around the globe. In this way, the book would emphasize that problematic border practices are not isolated cases, limited to a few locations, but that they are a systemic problem in a global order that divides the surface of the earth and its population into territorial nation states. While I tried to follow this advice as much as possible, I had to balance it with the need for scholarly rigor; otherwise, the credibility of my argument would have been compromised. In the end, I chose to illustrate my argument with examples that represent my areas of scholarly expertise, which disproportionately includes contexts from Europe, North America, and the global north. I encourage other scholars to examine whether and how my argument applies to different geographical and historical contexts.

Some readers may find that *Migration Borders Freedom* represents a Eurocentric and Western perspective in another way: my treatment of central concepts – such as the concepts of freedom and utopia – is located firmly in a European and Western philosophical tradition. These readers have a valid point. In this sense, this book may not offer a truly global or timeless perspective. Rather, it recognizes that knowledge about freedom, borders, and migration is not universal but always situated in particular geographical and historical contexts.

Acknowledgments

I completed the manuscript of the book during a sabbatical year, which I spent at the Albert-Ludwigs University of Freiburg in Germany, where I enjoyed the hospitality of Tim Freytag and his colleagues at the Institute for Environmental Social Sciences and Geography. This stay was enabled by the Konrad Adenauer Research Award, which is granted jointly by the Alexander von Humboldt Foundation and the Royal Society of Canada. I also thank the Social Sciences and Humanities Research Council of Canada, which supported the research for other parts of the book. The index of this book was funded by a grant provided by the office of the Dean of Arts, Ryerson University.

At Ryerson University, I thank my colleagues in the Faculty of Arts and the Faculty of Community Services, in particular the Department of Geography and Environmental Studies, the Graduate Program for Immigration and Settlement Studies, and the Ryerson Centre for Immigration and Settlement, for providing a supportive, dynamic, and highly stimulating environment. The individuals who deserve special mention include Jean-Paul Boudreau, Wendy Cukier, Usha George, Janet Lum, Claus Rinner, John Shields, Myer Siemiatycki, Vappu Tyyskä, and Shuguang Wang. I also thank Ratna Omidvar and her team at the Global Diversity Exchange for their inspiring work linking research with practice.

While I was in Germany, I had enriching discussions with Heike Drotbohm, Anna Lipphardt, Albert Scherr, and Inga Schwarz and their colleagues at the Freiburger Netwerk für Migrationsforschung. Over the course of developing the material for this book, I benefited from discussions with many good colleagues and friends, including Bernd Belina, Uli Best, Ranjit Bhaskar, Franck Düvell, Salvatore Engel-di Mauro, Fabian Georgi, Ken Hewitt, Dan Hiebert, John Kannankulam, Audrey Kobayashi, Valerie Preston, Michael Samers, Nik Theodore, and Christina West. In addition, I received valuable feedback on various ideas presented in the book from John Agnew, Klaus Kufeld, Fiona McConnell, Alexander Murphy, Ulrike Ramming, Michael Weingarten, and Jan Winkler.

The research for the book has been an ongoing effort over a number of years, and during this time I benefited from the help of outstanding research assistants. In particular, Clair Ellis and Charity-Ann Hannan at Ryerson University, and Marc Schulze and Helge Piepenburg at the University of Freiburg were highly competent researchers who helped me identify and compile the information I needed to make the arguments presented in the book. Birgitt Gaida deserves special thanks for designing the two maps, and Michael Bauder (not related) for preparing the data for one of the maps.

Three anonymous reviewers provided excellent feedback on the book's proposal, which helped me strengthen many parts of the book. In addition, Sutapa Chattopadhyay and Pierpaolo Mudu commented on the Introduction; their insights and constructive critique challenged me to refine my argument throughout the book. Karen Uchic read the entire manuscript, provided competent advice on how to make the text accessible and readable, and should be credited for the idea to use the term "possibilia," which plays an important role in this book. At Routledge, I thank commissioning editor Faye Leerink, and her colleagues Emma Chappell, Priscilla Corbett, Cathy Hurren, Dawn Preston, and Megan Smith. The index was compiled by Sarah Ereira.

Many of the ideas used for this book developed over the past decade, and appeared in a series of previously published, peer-reviewed journal articles. My initial exploration of the topic of open borders led to the publication of the article "Equality, Justice and the Problem of the International Borders: A View from Canada" in the open-access journal *ACME: An International E-Journal of Critical Geographies* (volume 2, issue 2, pp. 167–82) in 2003. The editors of the journal at the time, Lawrence Berg, Caroline Desbiens, and Pamela Moss, thought that this topic was worthy of inviting Uli Best, Franck Düvell, Dan Hiebert, Valerie Preston, and Michael Samers to critically comment on my ideas. The papers were published together in a themed section "Engagement: Borders and Immigration." This interest in the topic of open borders motivated me to pursue the topic further. Over the subsequent years, my search for a better understanding of the problem of border controls and its potential solutions led me to expand the focus of my research to the fields of critical borders studies, citizenship studies, and no-border perspectives, and to concepts such as domicile and urban citizenships, the right to the city, utopia, and the possible.

Over the years, I published several journal articles on these topics. It is the very nature of a journal article to make a narrow, single point, and this book gave me the opportunity to revisit some of those articles and develop them into a more far-reaching and cohesive narrative. For inclusion in *Migration Borders Freedom*, I revised a number of publications significantly from their original form and added previously unpublished material. I thank the journals for letting me reuse the material from the following single-authored articles in revised form:

"Perspectives of Open Borders and No Border," *Geography Compass* 9 (7) (2015): 395–405.

"Possibilities of Open Borders and No Border," *Social Justice* 39 (4) (2014): 76–96.

"Open Borders: A Utopia?" [Un monde sans frontières: une utopie? translation: Sophie Didier], *Justice Spatiale/Spatial Justice* 5 (December) (2013): 1–13 (available at: http://www.jssj.org/article/un-monde-sans-frontieres/).

"*Jus domicile*: In Pursuit of a Citizenship of Equality and Social Justice," *Journal of International Political Theory* 8 (1–2) (2012): 184–96.

"Towards a Critical Geography of the Border: Engaging the Dialectic of Practice and Meaning," *Annals of the Association of American Geographers* 101 (5) (2011): 1126–39.

"Possibilities of Urban Belonging," *Antipode* 48 (2) (2016): 252–71.

1 Introduction

> What we seek is freedom. Freedom to move, return, and stay
>
> Syed Khalid Hussan (2013, 280)

> *A drowned two-year-old boy became the first known migrant casualty of the year on Saturday after the crowded dinghy he was travelling in slammed into rocks off Greece's Agathonisi island.*

This is how the consortium the "Migrants' Files" recorded the tragic and unnecessary death of a toddler on January 2, 2016. The consortium, created by journalists from over 15 countries in Europe, aims to provide reliable and comprehensive data about the men, women, and children who have perished during their attempt to reach Europe. It does more than count the dead: it gives the migrants a human face by recording their names, age, gender, and the exact location where they died or went missing (Migrants' Files 2016).

A small sample of the entries recorded on the consortium's website paints a grim picture. On December 24, 2015, while the Christian world prepared to celebrate the birth of its savior, "at least 18 migrants drowned when their overcrowded boat sank in the Aegean Sea, the Turkish coastguard recovered the bodies including several children from the sea, and were hunting for another two who were missing." On August 27, 2015: "Up to 200 bodies have been discovered floating off the coast of Libya." Earlier in the year "about 400 migrants are feared died in an attempt to reach Italy from Libya when their boat capsized, survivors said" (April 13, 2015). "An overcrowded boat broke apart shortly after leaving Tripoli en route to Italy," resulting in 600 missing people who likely drowned (May 8, 2011). The list goes on and on. It includes not only drownings, but also fatalities among stowaways in trucks, deaths by starvation and exhaustion, migrants shot to death by border guards, desperate suicides, and other causes of death. A map of the 196 recorded incidents in 2015, resulting in 1,472 deaths and 2,130 persons missing, shows that attempts to enter Europe by sea claimed the most human lives (Figure 1.1). Most fatalities occurred in the Aegean Sea as migrants tried to reach Greece from Turkey, and off the coast of Libya as they attempted to reach Italian shores. Border-related deaths also occurred after migrants had crossed the

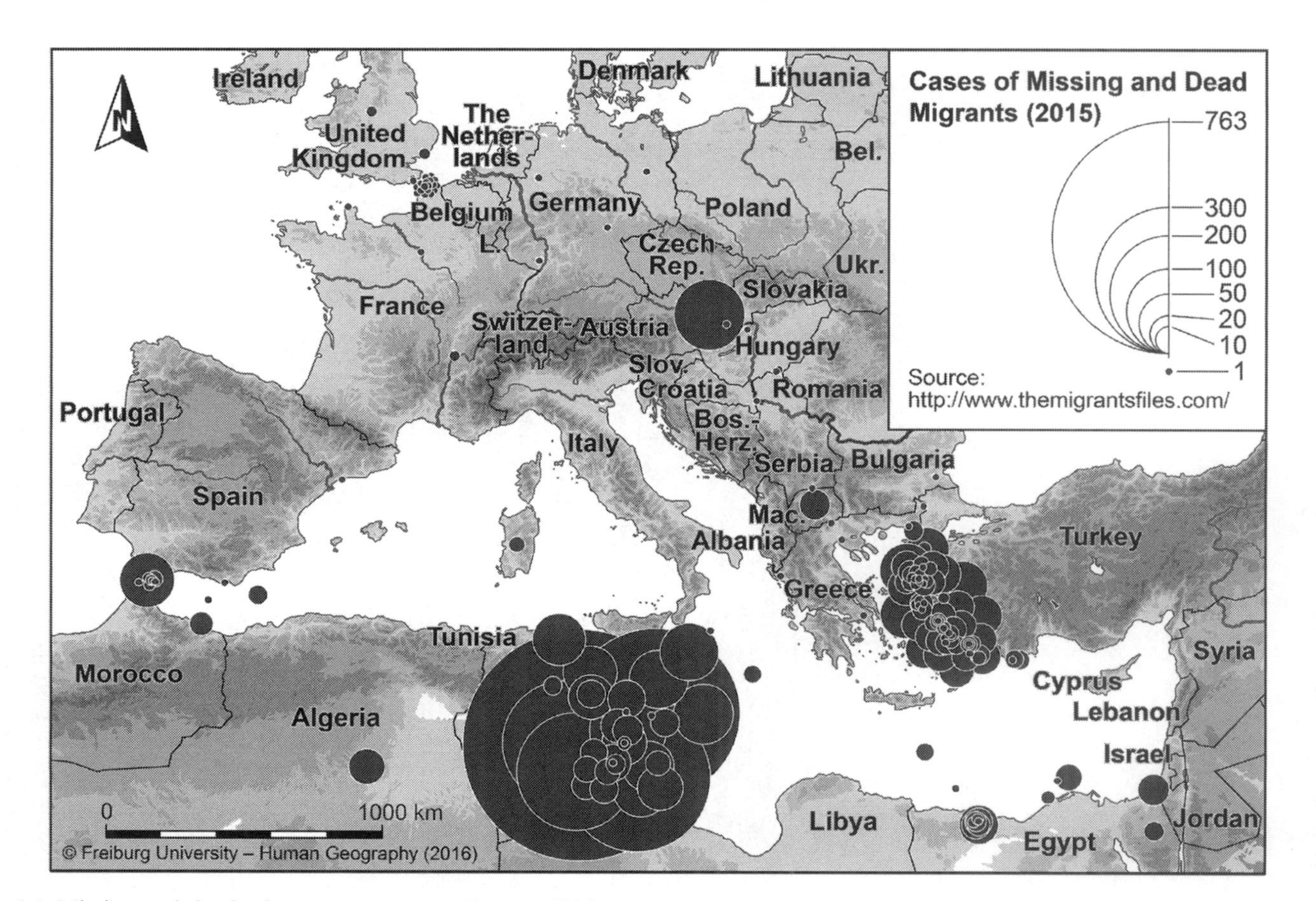

Figure 1.1 Missing and dead migrants en route to Europe, 2015
Source: Migrants' Files; map by Birgitt Gaida

physical border, such as in Austria, where 71 migrants were found dead in the back of a truck in August of 2015. Altogether, the database contained 3,049 entries in early January 2016, with an estimated 31,811 men, women, and children dead or missing since 2000. The numbers are staggering. The disheartening truth, however, is that the actual numbers are even higher. Despite the journalists' valiant efforts to record carefully every fatality, many deaths occur that nobody sees or documents.

In Australia, researchers at Monash University have created a similar database. The Australian Border Deaths Database records the known deaths resulting from Australia's border practices. It contains entries such as the drowning of 58 persons on April 11, 2013, including "Rehmatullah Muhammad Kan, male; Mahidi Fidayee, 16 years, male; Abdul Aziz, 63 years, male; Ibar Hussain Rajabi, 17 years, male; the rest unknown, all Afghan." Their boat was "lost at sea in Sundra Strait off Indonesian coast carrying 72 asylum seekers bound for Australia. 14 survivors found, 5 confirmed deaths, 53 missing presumed drowned." Another incident was the loss of 353 persons on October 19, 2001 – 146 children, 142 women, 65 men from Iraq and Afghanistan – who "drowned after [their] refugee vessel codenamed 'SIEV X' sank off Indonesia, but in Australian aerial border protection surveillance zone." Altogether, the database recorded 1,947 deaths between early 2000 and January 2016 (Border Crossing Observatory 2016). As in the case of the European statistics, the actual number is likely much higher (Pickering and Cochrane 2012).

Meanwhile, in the USA, forensic anthropologist Lori Baker is running a lab at Baylor University in Texas, where she and a team of scientists and students extract and analyze the DNA from the remains of migrants who died trying to cross the border from Mexico into the USA. Baker told the *Los Angeles Times* about her first case: in 2003 she examined the bones of a woman found in 2003 in Pima County, Arizona. A voter registration card found nearby provided clues about the identity of the deceased migrant, and Baker's analysis revealed that the DNA and the name on the card matched.

> Rosa Cano Dominguez, 32, was a mother of two from the Yucatan region who had been traveling to work in the Pacific Northwest when she sprained her ankle. She was abandoned by smugglers.
>
> (Hennessy-Fiske 2013)

The scientist and the migrant had a lot in common. Both were pregnant, working mothers in their 30s, and both were from families with low socioeconomic status. "I cried and cried over that case," Baker told the *Los Angeles Times* reporter, revealing the emotions she experienced when she discovered who the deceased person actually was.

Deaths at the border occur not only at the perimeter of rich countries in the Global North. In 2015, the global media reported about thousands of Rohingya people who were rescued at sea by Southeast Asian fishermen after fleeing Myanmar, where they have been denied citizenship and faced various forms of abuse. Since legal ways to migrate were unavailable and neighboring countries said they would not take them, the Rohingya were forced to rely on unscrupulous smuggler syndicates. These smugglers then abandoned the refugees at sea, often leaving them without water or food (NPR 2015). If they succeeded in bringing the refugees to their agreed-upon destination, they often held them to ransom in jungle camps to extract additional funds from them or their families. The media reported about mass graves littering the border between Thailand and Malaysia, containing the bodies of Rohingya who did not survive the brutal conditions in the camps or who were killed outright (Davis and Cronau 2015; Beech and Kelian 2015).

International borders have become deadly barriers that are on a par with war, genocide, and major epidemics and natural disasters in the number of fatalities they produce (Brian and Laczko 2014). Although border deaths are not a recent phenomenon, the horrific death counts of migrants in the Mediterranean Sea, in the waters between South East Asia and Australia, along the US–Mexico border, and in the waters of Southeast Asia illustrate the catastrophic dimensions this phenomenon has now assumed.

Migration scholars are increasingly speaking of "border regimes" to capture the complex and ever changing practices that govern migration (Tsianos and Karakayali 2010). These regimes do not neatly distinguish between the migrant-as-victim and the state seeking to constrain mobility. Rather, they focus on the interplay between governments and administrations, civic institutions, other actors using various technologies of surveillance and mechanisms of control, and migrants' efforts and motivations to circumvent these technologies and mechanisms. These regimes have assumed a new quality that explains a large portion of the increase of migrant deaths in recent years.

These deaths are not primarily a problem of smuggling, as the mainstream politicians and media would like us to believe. Certainly, there are smugglers without scruples who care little about migrants' lives and use the migrants' vulnerable situation to extort as much money from them as possible. These smugglers are monsters. Without closed borders, however, these smugglers would not have any desperate "customers" to prey upon.

People have always migrated to escape war and hunger, to be with loved ones, or to seek out greener pastures. Today, however, advancements in transportation have made travel faster and cheaper, and communication technologies have made it possible to connect with family and friends independent of physical distance. As a result, the mobility of the global population has increased in volume and migration flows have diversified. At the same time, political developments have fostered global migration. With the downfall of the Iron Curtain, for example, migration has become possible for significant numbers of people in Asia and Europe; and the political turbulences following the Arab Spring displaced millions of people and forced them to cross borders to seek refuge. Meanwhile, a parallel development is the increasing integration of national economies and a corresponding growing political interdependency of countries.

The growing global political and economic integration prompted globalization scholars and corporate strategists to predict that borders would become irrelevant (e.g. Ohmae 1991, 1995). From today's vantage point, these predictions were wrong. Rather than vanishing, borders continue to be highly relevant. Faster and cheaper transportation does little good when entry to safer destinations in Europe, Australia, the USA, and other countries is off limits. In fact, many people are completely immobilized when they are detained in their attempts to cross these borders. In the context of migration, borders and their regimes are not disappearing but are becoming stronger and increasingly deadly.

Interestingly, the enduring relevance of borders has gone hand in hand with the ongoing transformation of economic and political relationships between nation states. Europe exemplifies how these relationships are in constant flux: the European nation states that belong to the Schengen Area may have opened their borders to each other's citizens, but, at the same time, they militarized the border at the perimeter of the Schengen Area. In 2015, some nation states, such as Austria, Germany, and Sweden, temporarily re-established controls at their borders in an effort to regulate the migration of refugees. The continual changes in policies and practices related to migration give me hope that the hardening of borders for a large part of the world's population is not an unstoppable trend that will end only when borders are completely sealed. Rather, governments and other actors involved in regulating migration may come to their senses and realize that it is impossible to completely seal borders; they may seek alternative solutions by reducing the barriers to migration for everyone and, at one point, eliminate them altogether.

In the long run, political and economic structures are likely to continue breaking away from the national scale. Although the national imagination is still a powerful force – for example, to mobilize national electorates, as can be seen by the recent rise of nationalist anti-immigrant political parties and party programs throughout the Global North – it may eventually be replaced by new geopolitical imaginaries. Sociologist Saskia Sassen (2008, 147) sees "globalization and electronic networks" as such new imaginaries that will sooner or later transform politics and rearrange political systems as we know them today. If indeed this trend continues, the long-term scenario in which borders are irrelevant may be possible after all. People may then be free to migrate.

Currently, however, migration is still controlled and negotiated at international borders and by national citizenship, which most people receive at birth. In fact, free cross-border mobility would do little to improve the situation of migrants if they continue to be put in danger due to lack of citizenship. To address the root problem of closed borders and exclusion, we need to ask ourselves some tough questions. Should migration be unconstrained by international borders? How can such firmly established political practices and principles regarding national boundaries change? What kind of political imaginations would be required for a world in which all people possess the freedom of migration? *Migration Borders Freedom* seeks to answer these questions.

Freedom, Borders, Migration

Modern society cannot be imagined without the concept of freedom. It was central to the philosophies of enlightenment thinkers, such as Immanuel Kant and Georg W. F. Hegel, as well as John Locke and Adam Smith. Their ideas shaped not only the field of philosophy but also the political and economic systems that organize almost every aspect of our lives today. Nevertheless, there is no universally accepted definition of freedom. Instead, this concept has been interpreted in various ways.

One interpretation of the concept of freedom relates to the autonomy of individuals to reason and decide. This autonomy includes the freedom to decide on religious matters, freedom of speech, the freedom to negotiate and sign a contract, to buy, sell, and own property, and the freedom from being told by others what to do and how to live. This liberal interpretation of freedom relates to the concept of equality: every person should equally be able to enjoy freedom, and no person or group should possess an asymmetrical ability or right to interfere with another person's freedom. Individual freedoms have also provided the philosophical basis for a set of laissez-faire economic and political practices commonly known as neoliberalism, which emphasizes the freedom to own, trade, and use property and the freedom of contract.

According to critics, the liberal interpretation of freedom is an ideological deception. Half a century ago, philosopher Herbert Marcuse (1964) observed that mass society had appropriated the vocabulary of the enlightenment, including the concepts of freedom and equality, and puts this vocabulary to use in a way that constrains rather than enables independent thinking and human emancipation. More than four decades later, geographer David Harvey made a similar observation when he examined the history of neoliberalism. According to Harvey (2009, 2005, 5–38), neoliberal ideology has applied the concept of freedom in a rather narrow sense to market forces, enterprise, and property ownership. This particular application justified the expansion of capitalist practices into ever more aspects of our lives and into the last remaining corners on the surface of the earth. In the name of freedom, societies have commodified water resources and ecologically sensitive woodlands, scientific knowledge, genetic codes of living organisms, and the care for their children and elderly. Even air space is for sale. My own employer, Ryerson University, traded the rights to the "air" above its parking structure to allow a private company to build a movie theater in exchange for using the facility as a lecture hall during the daytime. A bewildered university president, Sheldon Levy, quipped: "Who knew about air rights?" (Brown 2015), expressing his surprise about his university's "freedom" to rent out even the air above it.

However, the liberal interpretation of freedom is not the only one, or even the most compelling. Indeed, the liberal and neoliberal incarnations of individual freedom have often restrained other types of freedom, such as the freedom of self-determination, the freedom from subordination and domination, or the freedom from exploitation and unfair distribution of wealth and opportunity.

Harvey, for example, notes that the "values of individual freedom and social justice are not … necessarily compatible" (Harvey 2005, 41). A philosophical tradition that traces its origins to Georg W. F. Hegel and Karl Marx produced a narrative of freedom that relates to social justice. According to this narrative, subordinated and precarious groups are entitled to freedom from exploitation and oppression by prevailing political and economic structures. This structural understanding of freedom often conflicts with freedoms that emphasize the individual.

There are still other interpretations of the concept of freedom. The historian Michel Foucault sees freedom "as an element that has become indispensable to governmentality" (Foucault 2007, 353). He points to a "complex interplay" between freedom and power, one necessitating the other (Foucault 2002, 342). According to Foucault, freedom and power cannot be neatly separated into two antagonistic forces.

The political theorist Hannah Arendt (1960) conceptualizes freedom in yet another way: she differentiates between the concepts of freedom and free will. She associates free will with the intrinsic human capacity to make autonomous decisions about available options. Freedom, on the other hand, is the capacity to begin something new: the ability "to call something into being which did not exist before, which was not given, not even as an object of cognition or imagination, and which strictly speaking could not be known" (Arendt 1960, 32).

Arendt, Foucault, and Harvey seem to agree that freedom should not be equated with the ability to retreat from politics or to live free from political interference. Rather, freedom is an inherently political concept that requires interaction with others and "worldly space to make its appearance" (Arendt 1960, 30). Following Arendt, freedom is achieved through action – not to attain a preconceived goal but to create our own future through transformative social and political practice.

The diverse ways of understanding the concept of freedom – as individual autonomy to decide, human equality, absence of structural oppression, and the capacity to create one's future – will reappear throughout this book. Given the various ways to interpret this concept, let me state my position from which I develop my argument. My starting position is simple: all human beings possess the freedom of migration and they should be able to exercise this freedom.

This starting position is not so far-fetched. Hannah Arendt, in accepting the Lessing Prize of the Free City of Hamburg, said: "Of all the specific liberties which may come into mind when we hear the word 'freedom,' freedom of movement is historically the oldest and also the most elementary." Since ancient times, restriction on the freedom of movement has been a condition of enslavement. Important for my purposes is Arendt's observation that "freedom of movement is also the indispensable precondition for action" (Arendt 1968, 9) and thus a requirement for people to achieve political and social transformation. Without freedom of movement, people cannot create their own destiny. To deny a person the freedom of movement infringes on a person's ability to participate in transformative politics. In other words, freedom of movement is central to human liberation.

That freedom of mobility is related to other types of freedom has not been lost on social and political activists, such as Syed Khalid Hussan, whose epigraph opens this chapter. According to social-justice activist Harsha Walia (2013, 77) "the freedom to stay and resist systematic displacement, the freedom to move in order to flourish with dignity and equality, and the freedom to return to dispossessed lands and homes" is fundamental to the liberation from destructive capitalist practices, racism, colonialism, and other forms of oppression. Nowhere are the consequences of constraining the freedom of movement more apparent than when people are denied crossing international borders, when they die in the attempt to do so, or when they are treated in demeaning and dehumanizing ways simply because they crossed an international border.

That sovereign states claim a monopoly over the mobility of people across state borders puts a dampener on the prospect of human freedom. In today's world, in which the human population is "divided up into mutually exclusive bodies of citizens, international migration is an anomaly with which the state system has some awkwardness coping," says sociologist John Torpey (2000, 123). The typical response by states is to prevent free cross-border migration. If necessary, "people with guns are prepared to enforce the boundaries" (Carens 1995, 2). Attempts to evade these guns can put the migrants at great risk, resulting in the horrific death counts that I described earlier.

Liberal political thinkers would counter that constraints to individual freedom are justified by the democratic process. In sovereign democratic countries, such as the United States, it is up to the American people to decide who they permit to cross their border and live within their territorial boundaries. Aside from a deficit of democracy that tends to muffle the voices of disadvantaged and racialized groups, women, and especially Indigenous peoples within the USA, the logic of this liberal political argument is highly problematic because it only applies to the population inside the USA, but not the people living outside of it. In fact, borders are notoriously undemocratic. Democracy entails that people affected by decisions should be involved in making them. In the context of cross-border migration, only the people on one side of the border are included in the decision-making process; the migrants on the other side of the border are excluded, although they are often more affected by these decisions than the decision makers. The constraints on freedom of migration tend to be blatantly asymmetrical, and vary depending on which side of the border a person is located.

Imagining Freedom of Migration

The discussion about the concept of freedom gives me the opportunity to introduce other themes and the corresponding vocabulary that weave through the book. One theme is that ideas and concepts do not exist in isolation from worldly context. For example, the quest for freedom arises when people experience unfreedom. Most societies possess many freedoms that are never perceived as freedoms, such the freedom to breathe (Hawel 2006). Imagine a scenario in which all available air has become a tradable commodity that has become rare due to ongoing pollution so that only the rich can afford it. In such a scenario people will become well aware how pollution and the commodification of air infringe on their freedom to breathe. The right to freedom of breathing is a non-issue when everyone has the ability to act on it. But when that right is denied, it becomes a problem.

Just as they possess the freedom to breathe, all human beings possess the freedom of migration, even if they are not aware that they possess it. The demand to act on this freedom arises when people need to escape violence, hunger, or oppression, or to create their own future. We realize that this freedom has been denied when people drown, die from starvation and exhaustion, are shot, or left to die in the desert during their attempts to exercise their freedom to migrate. Thus, the freedom to migrate becomes necessary only as a result of history or politics. It becomes a problem worth contemplating when it is absent or constrained to a degree that it causes suffering and death; otherwise, it is a non-issue. It simply doesn't exist. In this way, freedom – including the freedom of migration – is an inherently dialectical concept.

Ah, dialectics – a cringe-worthy term for the uninitiated. Not everyone seems to share my excitement for this term. At the risk of failing in my attempt to avoid academic jargon, I decided to hold on to the term because dialectics is a critical scientific tool to understand complex concepts, such as freedom or borders, and to develop solutions addressing problematic practices associated with these concepts. Due to its importance to my argument in this book, dialectics weave like a common thread through its pages.

One of the core ideas of dialectical thinking is that the world is full of contradictions, which we must address head on rather than brush aside as inconveniences. In fact, freedom, as a dialectical concept, embodies "a whale of a contradiction," says David Harvey (2014, 203). A fundamental contradiction is that "freedom and domination go hand in hand. There is no such thing as freedom that does not in some way have to deal in the dark arts of domination." Domination can occur by force. In antiquity, defeated enemies were forced into slavery so that the victors could enjoy freedom from labor. Today, the more likely scenario is that domination is enforced through "ideological manipulation" (2014, 204). In this context, the asymmetrical distribution of freedom and domination among people is justified under the guise of societal consent or the supposedly natural forces of the market. Harvey concludes: "Clearly at the root of the dilemma lies the meaning of freedom itself … It is impossible to escape the contradictory unity of freedom and domination no matter what politics are espoused" (2014, 206). A dialectical approach enables me to face head on the contradictions embodied in such concepts as freedom and the "border."

The dialectical approach is also useful in thinking about possible paths towards addressing some of today's problematic border practices that infringe on the freedom of migration. In the same way as we can plan a journey that leads us through mountainous terrain or that provides us with ocean views, we can envision paths towards the future that emphasize various aspects of freedom and ideas of belonging and living together. Moreover, we can embark on a journey that ends in the neighboring village or at a distant and unknown location that lies beyond the horizon, behind the mountain range or ocean where we cannot see. In the same way, our imagination can focus on practical and feasible alternatives, or a distant future that cannot yet be grasped.

Pondering possible paths towards a future in which humanity can exercise freedom of migration is an important aim of *Migration Borders Freedom*. Harvey (1972, 11) called upon geographers more than 35 years ago to "formulate concepts and categories, theories and arguments, which we can apply in the process of bringing about a humanizing change." Drawing on the work of Harvey and other scholars, such as Theodor W. Adorno, Ernst Bloch, and Henri Lefebvre, I explore possibilities ranging from the practical and feasible to the distant and utopian. On the one hand, there are alternatives within our reach that would allow people to migrate freely across national borders and belong to the communities in which they arrive. On the other hand, there are more far-reaching possibilities that escape our imagination because the contexts in which these freedoms unfold do not yet exist. In reference to Adorno's work, Marcus Hawel (2006, 105, my translation) explains that "the idea of a liberated society is necessarily a negative utopia." It is not a condition that we can envision from our contemporary vantage point. Rather, freedom emerges when social and political practices engage and transform the conditions that have produced unfreedom.

Utopian imaginaries have become rare in scholarly and public debate. Once they were a staple in forward-looking scholarship, activism, and politics, and inspired practitioners to translate these utopias into practice. They illustrated, for example, how people could live in harmony with nature and inspired city planners to build parks and green space. Unfortunately, utopia was also associated with Soviet-style socialism and its grand vision of communism. With the Soviet Union's fall, utopia was also dismissed as an idea that serves to enslave rather than liberate people. While I do not mourn the rejection of utopia as a grand vision of an alternative world, I nevertheless feel inspired by the utopian possibility of achieving a world in which we all possess freedom of migration.

The utopian possibility is especially important when the freedom of migration is denied to people. In the context of the *sans-papiers* in France and the criminalization of migrants elsewhere, sociologist Pierre Bourdieu, in a conversation with Nobel literature laureate, Günter Grass, called upon fellow intellectuals to live up to their responsibility "to restore a sense of utopian possibility" (Grass and Bourdieu 2002, 66) that challenges the conditions that deny people their freedoms and liberties. *Migration Borders Freedom* is my attempt to answer this call.

Structure and Context of the Book

Migration Borders Freedom is divided into two parts, each part containing three chapters. The chapters of Part I represent a diagnosis of the practices and the corresponding ways in which people are making sense of borders and migration. The chapters of Part II formulate possible solutions that move beyond the current conditions. In other words, the narrative of the book advances from an analysis to a normative discussion of borders, migration, and freedom.

It is worth noting that the two parts of the book are thematically distinct. While Part I focuses on cross-border migration, Part II emphasizes citizenship and belonging. Although I have separated these themes in the way I organized the chapters, I recognize that "freedom of movement and freedom to inhabit are necessarily connected" (Loyd et al. 2012b, 10). In fact, one of the main points of the book is that freedom to migrate cannot be divorced from discussions of the right to stay and belong.

Migration Borders Freedom follows in a line of extraordinary books that were published over the last quarter century on the topics of borders, migration, and belonging. Path-breaking research has explored how borders and citizenship serve to regulate populations in light of globalization and migration (e.g. Bauböck 1994; Mau et al. 2012). Several books have also examined the economic, social, and ethical implications of greater mobility across national borders (Pécoud and de Guchteneire 2007; Schwartz 1995; Barry and Goodin 1992; Ghosh 2000a). Many of these books presuppose that nation states as we know them today persist, and then consider the consequences of migration from various perspectives, ranging from the theoretical to the empirical, and from the philosophical to the political and economic. They generally conclude that a multilateral approach is necessary to solve the problems created by today's border and migration practices. Senior Consultant to the International Organization for Migration, Bimal Ghosh (2000b, 25), for example, advocates for a compromise solution of "regulated openness" that lies somewhere between the positions of completely sealed and completely open borders.

Migration Borders Freedom is also inspired by research in the field of critical border studies that have emerged over the last few decades, and by other scholarship that has applied a critical-theory lens to the study of migration (Albert et al. 2001; van Houtum et al. 2005). The authors of such work have critiqued existing politics and practices – for example, the imprisonment, detention, and deportation of migrants (Loyd et al. 2012a; De Genova and Peutz 2010) or birthright citizenship and property ownership (Stevens 2010) – in far-reaching ways. They also acknowledge the autonomy of migrants and their capacity to act politically (Mudu and Chattopadhyay 2016), challenging the binary distinctions between migrants and citizens, included and excluded, and "us and them" (Anderson 2013). This scholarship tends to conclude that fundamental social and political transformation would be necessary to solve today's problems associated with borders and migration.

Migration Borders Freedom bridges many of the various perspectives that these previous works pioneered. But rather than seeking a compromise solution or lowest common denominator between them, it follows a dialectical approach that not only connects the themes of borders, mobility, and belonging but also enables us to see the value of both practical solutions and far-reaching inspiration. In this way, the book contributes a fresh perspective to a growing body of critical scholarship on borders, mobility, and belonging, with the ultimate aim of charting a course towards human liberation.

References

Albert, Mathias, David Jacobson, and Yosef Lapid, eds. 2001. *Identities, Borders, Orders: Rethinking International Relations Theory.* Minneapolis, MN: University of Minnesota Press.

Anderson, Bridget. 2013. *Us and Them? The Dangerous Politics of Immigration Control.* Oxford: Oxford University Press.

Arendt, Hannah. 1960. "Freedom and Politics: A Lecture." *Chicago Review* 14(1): 28–46.

Arendt, Hannah. 1968. *Men in Dark Times.* San Diego, CA: Harvest Books.

Barry, Brian and Robert E. Goodin. 1992. *Free Movement: Ethical Considerations in the Transnational Migration of People and of Money.* New York: Harvester Wheatsheaf.

Bauböck, Rainer. 1994. *Transnational Citizenship: Membership and Rights in International Migration.* Aldershot: Edward Elgar.

Beech, Hannah and Wang Kelian. 2015. "Rohingya Survivors Speak of Their Ordeals as 139 Suspected Graves Are Found in Malaysia." *Time*, May 26. Accessed January 7, 2016. http://time.com/3895816/malaysia-human-trafficking-graves-rohingya/.

Border Crossing Observatory. 2016. "Australian Border Deaths Database. Monash University." Accessed January 5, 2016. http://artsonline.monash.edu.au/theborder crossingobservatory/publications/australian-border-deaths-database/.

Brian, Tara and Frank Laczko, eds. 2014. *Fatal Journeys Tracking Lives Lost during Migration.* Geneva: International Organization for Migration.

Brown, Louise. 2015. "Sheldon Levy leaving Ryerson, and Toronto, a Changed Place." *Toronto Star*, February 20. Accessed January 2, 2016. http://www.thestar.com/news/insight/2015/02/20/sheldon-levy-leaving-ryerson-and-toronto-a-changed-place.html.

Carens, Joseph. 1995. "Immigration, Welfare, and Justice." In *Justice in Immigration*, edited by Warren F. Schwartz, 1–17. Cambridge: Cambridge University Press.

Davis, Mark and Peter Cronau. 2015. "Migrant Crisis: Rohingya Refugees Buried in Mass Graves near Thailand Authorities, Survivor Says." *ABC News*, June 23. Accessed January 7, 2016. http://www.abc.net.au/news/2015-06-22/rohingyas-secret-graves-of-asias-forgotten-refugees/6561896.

De Genova, Nicholas and Nathalie Peutz, eds. 2010. *The Deportation Regime: Sovereignty, Space, and the Freedom of Movement.* Durham, NC: Duke University Press.

Foucault, Michel. 2002. "The Subject and Power." In *Power, Essential Works of Foucault 1954–1984, Volume 3*, edited by James D. Faubion, 326–348. London: Penguin.

Foucault, Michel. 2007. *Security, Territory, Population: Lectures at the Collège de France*, edited by Michel Senellart, translated by Grahma Burchell. New York: Picador.

Ghosh, Bimal, ed. 2000a. *Managing Migration: Time for a New International Regime.* Oxford: Oxford University Press.

Ghosh, Bimal. 2000b. "Towards a New International Regime for Orderly Movements of People." In *Managing Migration: Time for a New International Regime*, edited by Bimal Gosh, 6–26. Oxford: Oxford University Press.

Grass, Günter and Pierre Bourdieu. 2002. "Dialogue: The 'Progressive' Restoration." *New Left Review* 14: 76–77.

Harvey, David. 1972. "Revolutionary and Counter Revolutionary Theory in Geography and the Problem of Ghetto Formation." *Antipode* 4(2): 1–13.

Harvey, David. 2005. *A Brief History of Neoliberalism.* Oxford: Oxford University Press.

Harvey, David. 2009. *Cosmopolitanism and the Geographies of Freedom.* New York: Columbia University Press.

Harvey, David. 2014. *Seventeen Contradictions and the End of Capitalism.* Oxford: Oxford University Press.

Hawel, Marcus. 2006. "Negative Kritik und bestimmte Negation: Zur praktischen Seite der kritischen Theorie." In *Aufschrei der Utopie: Möglichkeiten einer anderen Welt*, edited by Marcus Hawel und Gregor Kritidis, 98–116. Hannover: Offizin-Verlag.

Hennessy-Fiske, Molly. 2013. "Effort to ID Immigrants' Corpses Is Gratifying—and Sad." *Los Angeles Times*, November 1. Accessed January 4, 2016. http://www.latimes.com/nation/la-na-c1-baylor-bones-20131101-dto-htmlstory.html.

Hussan, Syed Khalid. 2013. "Epilogue." In *Undoing Border Imperialism*, Harsha Walia, 277–281. Oakland, CA: Ak Press.

Loyd, Jenna M., Matt Michelson, and Andrew Burridge, eds. 2012a. *Beyond Walls and Cages: Prisons, Borders, and Global Crisis.* Athens, GA: University of Georgia Press.

Loyd, Jenna M., Matt Michelson, and Andrew Burridge. 2012b. "Introduction." In *Beyond Walls and Cages: Prisons, Borders, and Global Crisis*, edited by Jenna M. Loyd, Matt Michelson, and Andrew Burridge, 1–15. Athens, GA: University of Georgia Press.

Marcuse, Herbert. 1964. *One-Dimensional Man: Studies in the Ideology of Advanced Industrial Society.* Boston, MA: Beacon Press.

Mau, Steffen, Heike Brabandt, Lena Laube, and Christof Roos. 2012. *Liberal States and the Freedom of Movement: Selective Borders, Unequal Mobility.* Basingstoke: Palgrave Macmillan.

The Migrants' Files. 2016. "The Human and Financial Cost of 15 Years of Fortress Europe." Accessed January 4, 2016. www.themigrantsfiles.com.

Mudu, Pierpaolo and Sutapa Chattopadhyay. 2016. *Migrations, Squatting and Radical Autonomy.* London: Routledge.

NPR. 2015. "Rohingya Migrants Left out at Sea, No Country Will Allow Them Ashore." Last modified May 20. Accessed January 7, 2016. http://www.npr.org/2015/05/18/407619687/rohingya-migrants-left-out-at-sea-no-country-will-allow-them-ashore.

Ohmae, Kenichi. 1991. *The Borderless World: Power and Strategy in the Interlinked Economy.* London: Fontana.

Ohmae, Kenichi. 1995. *The End of the Nation State: The Rise of Regional Economies.* London: HarperCollins.

Pécoud, Antoine and Paul de Guchteneire, eds. 2007. *Migration without Borders: Essays on the Free Movement of People.* New York: Berghahn Books.

Pickering, Sharon and Brandy Cochrane. 2012. "Irregular Border-Crossings Deaths and Gender: Where, How and Why Women Die Crossing Borders." *Theoretical Criminology* 17(1): 27–48.

Sassen, Saskia. 2008. *Territory, Authority, Rights: From Medieval to Global Assemblages*, updated edition. Princeton, NJ: Princeton University Press.

Schwartz, Warren F., ed. 1995. *Justice in Immigration.* Cambridge: Cambridge University Press.

Stevens, Jacqueline. 2010. *States without Nations: Citizenship for Mortals.* New York: Columbia University Press.

Torpey, John. 2000. *The Invention of the Passport: Surveillance, Citizenship and the State*. Cambridge: Cambridge University Press.

Tsianos, Vassilis and Serhat Karakayali. 2010. "Transnational Migration and the Emergence of the European Border Regime: An Ethnographic Analysis." *European Journal of Social Theory* 13(3): 373–387.

van Houtum, Henk, Olivier Kramsch, and Wolfgang Zierhofer, eds. 2005. *Bordering Space*. Farnham: Ashgate.

Walia, Harsha. 2013. *Undoing Border Imperialism*. Oakland, CA: AK Press.

Part I
Diagnosis

> When philosophy paints its grey in grey, then a configuration of life has grown old, and cannot be rejuvenated by this grey in grey, but only understood; the Owl of Minerva takes flight only as dusk begins to fall.
>
> Georg Wilhelm Friedrich Hegel (1970 [1820], 59–60)

> The philosophers have only interpreted the world in various ways; the point, however, is to change it.
>
> Karl Marx (1964 [1845])

The freedom of migration stops for the majority of the global population at the border. In this part of the book, I offer a diagnosis of border practices that create problems – often deadly ones – for many migrants. I start by problematizing the very concept of the border. In Chapter 2, I illustrate how the single concept of the border can be understood in very different ways, depending on the particular purposes that borders serve. This chapter sets the tone for the remainder of the book in several ways: first, it shows that concepts such as the border do not possess a single and universal meaning, and that we must relinquish the idea that we can uncover such a meaning. Second, the chapter shows that context matters; depending on the situation in which people experience borders, the concept changes its meaning. Third, it introduces the reader further to the dialectical way of thinking. This way of thinking connects worldly contexts with the manner in which we understand the world. The dialectical way of thinking also enables us to grapple with the contradictions that these different understandings raise.

Chapter 3 continues in the spirit of this dialectical approach. This chapter shows how advocates for open borders have assumed multiple – often contradictory – philosophical positions to argue for the freedom of migration for all. These advocates argue for open borders for very different reasons. Nevertheless, our inability to force the calls for open borders into a single framework indicates that the path to open borders does not take us on a straight highway but on a rambling road with twisting turns and unexpected forks.

In the final chapter of Part I, Chapter 4, I investigate how we may envision a world of freedom of migration. While critiques of current border practices abound, concrete visions of a world of unconstrained migration across international borders are rare. In this chapter, I distinguish between two possibilities of a world of free migration: one that assumes that international borders will continue to exist but that these borders are open, and the more visionary no-border project, which sees the existence of national borders as a snap-shot in history that will at one point be superseded by different social and political arrangements.

The three chapters of Part I advance from assessing how the border is envisioned in different contexts, towards examining various perspectives of free migration, and finally exploring different ways of thinking about scenarios that would enable freedom of migration. Thus, the chapters progress from looking backward at existing conditions and practices to looking forward towards future possibilities. This in turn mirrors a progression accomplished in the 19th century by the grand masters of dialectical thinking, Georg W. F. Hegel and Karl Marx. While Hegel's Owl of Minerva acquired wisdom only at dusk, after the events of the day had occurred, Marx realized – as he scribbled his Thesis on Feuerbach in his notepad – that scholarship has a role to play in projecting the future.

References

Hegel, Georg W. F. 1970 [1820]. *Grundlinien der Philosophie des Rechts oder Naturrecht und Staatswissenschaft im Grundrisse* [Contours of the philosophy of right or natural right]. Stuttgart: Reclam.

Marx, Karl. 1964 [1845]. "Thesen über Feuerbach" [Theses on Feuerbach]. *Marx-Engels Werke.* Band 3. Berlin: Dietz Verlag. Accessed February 2, 2006. www.mlwerke.de.

2 Borders in Perspective

> Borderwork is less and less something over which people have no control.
>
> Chris Rumford (2008, 10)

The border is a central concept in the debate of migration. It is also highly ambiguous. Over the period that I completed this book, borders and migration were among the central topics in global news. A small selection of articles published in the *New York Times* and the *Guardian* illustrates the multiple perspectives from which the news has approached the border in the context of migration. Although this selection shows that borders constrain freedom, it is not clear *whose* freedom they constrain.

The biggest topic in 2015 was the refugee "crisis" in Europe. The summer and fall of that year witnessed the "largest movement of people across Europe since World War II" (Surk and Lyman 2015). The crisis began when an increasing number of migrants chose the so-called Balkan route to reach Central Europe. On October 27, the *New York Times* reported that

> fresh fighting in Syria and growing fears of border closings are driving more migrants to undertake the treacherous trek.
>
> At the moment, the biggest crunch appears to be on the southern border of Slovenia, a small Alpine nation on the Adriatic Sea that has become the gateway to Europe for migrants since Hungary closed its border with Croatia on October 16.
>
> ...
>
> In the past ten days, 83,600 migrants have crossed into Slovenia, government officials said, while 57,981 have crossed from Slovenia into Austria, and 14,000 are waiting in government reception centers.
>
> (Surk and Lyman 2015)

When the exhausted refugees were stopped at the border on their way to their desired destination, some of them grew impatient, even violent. An official with the Slovenian Interior Ministry explained the violence in this way: "These people just want to move on, and when they are made to stop, they get nervous and extremely unhappy and then such incidents happen" (Surk and Lyman

2015). The refugees wanted to cross the border because the border brought them to safety and offered the prospect of a life beyond poverty and despair.

For those refugees who made it into Central Europe, however, the border did not always deliver on the promise of security and hope. On November 13, 2015, terrorists attacked Paris in the heart of Europe, killing 130 persons. The *New York Times* reported shortly thereafter about a refugee who fled Afghanistan to live in Austria. He broke down crying when he heard about the attacks. "This was happening in Afghanistan," he said. With his flight from Afghanistan, he had hoped to escape this type of terror. He told the *New York Times*: "I want to be safe … but if this happens here, where do I go? Right now, I think of my future and I'm scared" (Smale and Bradley 2015).

While Syrians, Iraqis, and Afghans cross the border to save their lives and livelihoods, governments seek to protect their nations from the supposed threat that these refugees bring. In particular, the attacks of Paris evoked concerns in Western countries that terrorists would be among the refugees. The *New York Times* reported that three days after the terrorist attacks in Paris, US border patrol agents apprehended "five Pakistanis and one Afghan" who attempted to cross the border south of Tucson, Arizona. A day later, "eight Syrians – two women and four children all from two families – presented themselves to the authorities in Laredo, [Texas], and asked for refuge in the United States." These incidents triggered fear among US federal law makers that Islamic "militants could be hiding among people fleeing the Syrian civil war and other conflicts" (Pérez-Peña 2015). As a result, these law makers voted on November 19 to suspend the admission of refugees from Syria and Iran.

An apparently unrelated event was also in the headlines in the summer of 2015: the scandal surrounding allegations of corruption in the world soccer governing body FIFA. The investigations into these allegations, however, drew the *New York Time*'s attention to the labor conditions of migrant workers who were building the stadiums and infrastructure for the 2022 FIFA World Cup in Qatar (Meier 2015). Two years earlier, the *Guardian* had described how migrant workers from Nepal "died at a rate of almost one a day in Qatar" and compared the working conditions in Qatar to modern-day slavery. The newspaper's own investigation uncovered evidence suggesting that many of the Nepalese migrants working in Qatar "face exploitation and abuses that amount to modern-day slavery, as defined by the International Labour Organisation" (Pattisson 2013). The *Guardian* explained that the Nepalese workers have accrued large debts to pay the recruitment agents who arranged the work in Qatar. "The obligation to repay these debts, combined with the non-payment of wages, confiscation of documents and inability of workers to leave their place of work, constitute forced labor, a form of modern-day slavery," the *Guardian* (Pattisson 2013) argued.

Migration and labor policies and their enforcement (or lack thereof) have enabled Qatar to create an exploitable labor force of foreigners that the country needs. In this way, Qatar uses its border as a mechanism to manage its economy and meet the labor demands of large-scale construction projects, like the

2022 FIFA World Cup. For the workers, crossing the same border often coincides with the beginning of life in modern-day slavery.

Generally speaking, borders constrain freedom. However, from the preceding examples, it is unclear whose freedom exactly borders constrain: the freedom of people to bring themselves to safety from war and start a life without despair? The freedom of nation states to protect themselves from perceived threats? The freedom of employers to use the production factor labor most effectively? The freedom of a monarchy to host a high-profile sporting event? Or the freedom of workers to receive fair wages and be treated as human beings?

Simply put, there is no universal perspective that summarizes the effect that borders exert. In this chapter, I explore the multidimensional character of borders. Critical border scholars, such as the geographers David Newman and Anssi Paasi (1998), among others (e.g. Johnson et al. 2011; Wastl-Walter 2011), have long realized that an ambiguous concept like the border can be approached from different angles, and that the "border" has different meanings depending on the vantage point one assumes. Philosopher Étienne Balibar has suggested that the border is *polysemic* in nature, by which he means that borders "do not have the same meanings for everyone" (2002, 81). For the professor or business executive, the border may represent an opportunity to learn about new scientific discoveries or expand into a new national product market, while the young, unemployed job seeker who is denied a visa or work permit experiences the border as a barrier to improve her livelihood.

While scholars concur that the border concept embodies multiple dimensions (or aspects), there is little agreement of how many of these dimensions there are. Some scholars, such as geographers Heather Nicol and Julian Minghi (2005, 681), distinguish between "two very different ways of understanding borders." Others perceive more than two dimensions of the border. Political scholar Malcolm Anderson (1996, 2–3) offers "four dimensions," the political economist Emmanuel Brunet-Jailly (2005, 645) four different analytical "lenses," and the sociologist Rob Shields (2006) a four-part ontology of the border. Although these scholars use different terms, they illustrate a similar phenomenon: borders can assume multiple characters and meanings. One could ask, how many aspects of the border can be empirically validated? However, it is not my interest or intention in this chapter to count and catalogue border aspects. Rather, I ask the following question: how shall we engage the border concept in light of its multidimensional character?

As a point of entry towards such a general approach, I will explore how the various meanings of the border that people form in their minds relate to the worldly ways in which they use and experience borders. For the Syrian family fleeing war, the border signifies a gateway to safety and a better life; for the law maker with the mandate to protect the nation, it is a place where threats to national security appear. This approach builds on existing critical border scholarship (van Houtum et al. 2005). In Balibar's (2002, 75) words, the border has no "essence": there is neither a uniform meaning of the border nor an objective character that can be attributed to borders independent of human interpretation. A hypothetical Archimedean vantage point, which assumes that the world can be "objectively" observed in an all-encompassing way, does not exist; and it is therefore impossible to produce authoritative knowledge of a concept such as the border (Haraway 1991; Rose 1997). Rather, the various meanings of the border are grounded in a diverse range of circumstances, practices, and experiences. This also means that the different meanings of the border are always context-particular, partial, and incomplete. In this chapter, I will discuss how various circumstances, practices, and experiences create different meanings of the border.

This chapter also gives me an opportunity to delve further into the idea of dialectics, which I discussed in the previous chapter. In particular, I invite readers to think dialectically about borders and migration. The way I apply dialectical thinking to the border concept draws on the work of Georg W. F. Hegel (e.g. 2005 [1807]) and others who subsequently contributed to the rich tradition of dialectics in Western thought. This work demonstrates that concepts tend to be unstable and must be continually rethought on the basis of the contradictions they embody. This "dialectical movement" applies especially to the border concept.

Dialectical thinking also entails that we treat the border neither solely as a worldly thing nor as a product of pure thought. Rather, when people use borders in certain ways, migrate across them, or experience them as impenetrable barriers, they shape the meanings we attribute to borders. Thus, migrants, activists, policy makers, and scholars are not passive bystanders but active participants in giving borders their meanings.

What is a Border?

I will use the term "aspect" to refer to different meanings of the border. This term conveys that meanings of a concept like the border are guided by both the observer's experiences and the manner in which the observer is situated in particular circumstances. Thus, the professor traveling to a conference experiences a different aspect of the border than the unemployed job seeker hoping to find work abroad. In a previous publication, I elaborated on the way in which Ludwig Wittgenstein uses the term "aspect," and how it can be linked to Hegel's dialectic (Bauder 2011). Here, I think, a discussion of these philosophical intricacies would distract from the main message about borders and migration that I wish to bring across.

As the earlier examples illustrate, neither the media nor critical border scholars seem to have any problems seeing different aspects of the border. In the following discussion, I illustrate how different aspects of the border are grounded in particular worldly circumstances, experiences, and practices. The different aspects I discuss represent by no means an exhaustive list of all possible aspects of the border (which is not the aim of this chapter). Rather they illustrate how different uses and experiences of borders produce different meanings.

Border as Line

The first aspect represents the border as a line in Cartesian space. This is the cartographer's view of the border, who draws lines delineating countries on a map. Figure 2.1 depicts this aspect visually as a line that separates the country of Namibia from neighboring Angola, Zambia, Botswana, and South Africa. To the west, this line follows the shore of the Atlantic Ocean, and to the north and south the Kunene, Okavango, Orange, and Zambezi rivers. To the east and parts of the north, it is a straight line, arbitrarily drawn along the 20th and 21st meridians (east) and between the 17th and 18th parallels (south). A citizen of Namibia can travel more than 1,000 miles from Katima Mulilo in the north-east of the country to Lüderitz in the south-west and is not considered an international migrant. Even if this person were fleeing from "well-founded fear of being persecuted" (1951 Refugee Convention), she would not be considered a refugee but an internally displaced person. However, when she moves a few miles north, she crosses the international border to Zambia and becomes an international migrant.

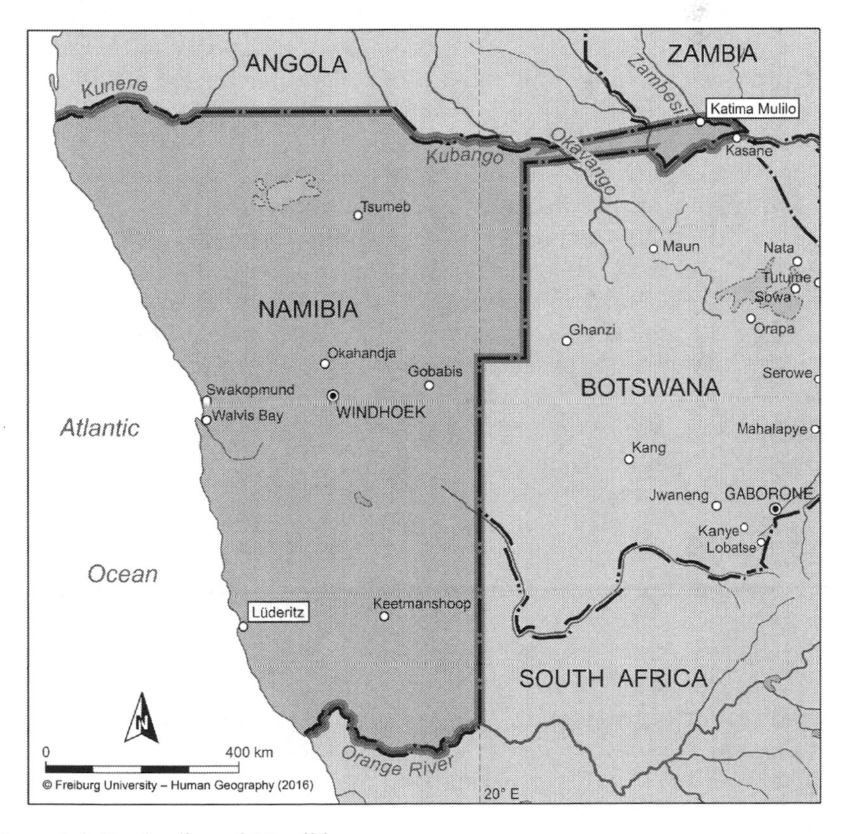

Figure 2.1 Border line of Namibia
Source: Map by Birgitt Gaida

The particular use of the border is critical to imagining it as a line in Cartesian space. In the case of many parts of Africa, the border lines that delineate the territories belonging to a state were introduced with European colonization. Where these lines were drawn was the result of negotiations between the competing European colonial powers and their interests in gaining geostrategic advantages, exploiting the continent's resources, and regulating commerce. These interests "inspired" them to draw arbitrary boundaries, often as a straight line on a map. For example, Namibia's northern border with Angola is the result of the 1886 boundary declaration between Germany and Portugal. Article 1 of this declaration specified:

> The Boundary line which shall separate the Portuguese and German Possessions in South-West Africa follows the course of the River Kunene [Rio Cunene] from its mouth to the waterfalls which are formed to the south of the Humbe by the Kunene breaking through the Serra Canna. From this point the line runs along the parallel of latitude to the River Kubango [Rio Cubango, Okavango], then along the course of that river to the village of Andara, which is to remain in the German sphere of influence, and from thence in a straight line eastwards to the rapids of Catima [Katima Mulilo Rapids], on the Zambesi [Zambezi].
>
> (quoted in *Geographer* 1972, 3)

The map of Namibia has an oddly shaped appendage (the German colonizers called it *Zipfel*) on the north-eastern corner, reaching into Botswana and Zambia. The earlier-mentioned town of Katima Mulilo is located in this appendage. This feature of Namibia's border line is another example of colonial border drawings. The German colonizers wanted access to the Zambezi river and therefore acquired this region in 1890 from Britain with the Heligoland-Zanzibar Treaty. They named the region the *Caprivizipfel* (Caprivi Strip), after Leo von Caprivi, who served as German Chancellor from 1890 to 1894. Only in 2013, was the former colonizer's name dropped and this region henceforth called the "Zambezi Region."

The commercial and geostrategic interests of the European colonizers disregarded the people who lived on the land. Correspondingly, the borders drawn by the Europeans did not consider the way the land was used by its residents. For example, the Kunene river that separates Angola and Namibia was a "location for communication" (Marx 2010) rather than a barrier. Drawing the border along this river divided, for example, the land used by the nomadic OvaHimba peoples. Throughout Africa, European colonizers ignored the geographical extent of linguistic, religious, and ethnic communities when they drew borders to delineate the territories they claimed. Sometimes they split communities; other times amalgamated antagonistic communities into a single state territory.

Today, the Cartesian border line is important to control migration. Some countries have erected walls and barbed wire fences along their borders to prevent people from crossing the border without authorization. One of the world's most infamous militarized borders exists along stretches of the United States' southern border to prevent unauthorized migration from Mexico. In the wake of the attacks of September 11, 2001, in New York and Washington, DC, and the subsequent "war on terror," the United States' northern border to Canada was also subjected to greater scrutiny of who crosses it. A problem emerged, however, when border officials were unable to locate the border because it was overgrown by forest. The line was no longer visible, which prompted a Canadian official of the International Boundary Commission to remark: "If you can't see the boundary, then you can't secure it" (Alberts 2006). In this case, the need to keep people from freely crossing the border requires the exact location of the border line to be visible on the earth's surface.

This aspect of the border as a line in two-dimensional space is, of course, limited and incomplete. The very geometry of the line dissolves as migration flows are increasingly monitored remotely at airports or transit hubs before migrants reach the actual border line, or at workplaces and in public spaces after they have crossed that line (Vaughan-Williams 2008). Balibar (1998, 217–18) observes that "borders are no longer at the border." Other border scholars are talking about the "externalization" of the border, suggesting that, for migrants, the border is no longer at the outer perimeter of a national territory but well outside of it, at locations where their intention to cross a border is assessed and, if necessary, prevented. In a similar way, one can speak of the "internalization" of the border as migrants are checked for their status after they have entered the territory of a country. A simple line on a map represents only a narrow and partial view that does not capture migrants' entire experience of the border.

Bastion of Sovereignty

Another aspect describes the border as an instrument of the state to exercise sovereignty. In her analyses of immigration law, legal scholar Catherine Dauvergne (2007, 2008) calls migration controls "the last bastion of sovereignty." In this case, the border is not represented as a line in Cartesian space but rather as a legal boundary that grants or denies access to the national community.

This aspect, too, is a product of history. In medieval Europe, before a modern territorial state emerged, migration was typically controlled by the prince, lord, or local authority to whom persons, families, or social groups were bonded. With the establishment of the "Westphalian" model, sovereign territorial states began monopolizing control over the mobility of people. In this way, these states tried to control the membership of their national communities. This process of the nation asserting control over migration

> took hundreds of years to come to fruition. It followed the shift of orientations from the local to the 'national' level that accompanied the development of 'national' states out of the panoply of empires and

> smaller city-states and principalities that dotted the map of early modern Europe.
>
> (Torpey 2000, 8)

This process was still ongoing in the early 20th century. For example, the British Aliens Act of 1905 presented a change away from the prevailing control of local authorities and agencies over migration. "This set of dispersed arrangements was now replaced by a policy of rejection, operated at the port of entry by central government … The history of immigration control was thus at the very front of a process of state formation" (Feldman 2003, 175).

The process of the sovereign territorial national state assuming control over migration could also be observed in the United States. Immigration had been the responsibility of the individual states through a variety of local laws and policies, such as preventing the arrival and entry of convicts in the 1780s (Neuman 2003; Zollberg 2003). This practice changed in the second half of the 19th century, when the economic opportunities that accompanied industrialization and the closing of the frontier attracted large numbers of immigrants to the United States. As a response to this development, the US federal government passed the 1875 Page Act and the Immigration Acts of 1882 and 1891, thereby asserting a stronger role in regulating immigration and who would subsequently become a citizen. Today, it is often taken for granted that immigration is national domain.

The border, in this context, is imagined as an instrument used by nation states to control their membership and protect political and civic order. This aspect of the border relates to formal citizenship through which states, in the words of John Torpey (2000), "embrace" their subjects. Of course, this aspect of the border, again, tells only a part of the story of how borders are used and experienced.

Labor Regulator

A different aspect emphasizes the effect borders have on labor. Many migrants experience the border as a mechanism that controls, disciplines, and in many cases exploits their labor. There are two ways in which borders do this: first, borders geographically divide the global workforce into countries with different labor and wage standards. An example is the border between Mexico and the United States. North of this border, in the United States, per capital gross national income was about US$55,200 in 2014; south of the border, in Mexico, it was only US$9,860 (World Bank 2015). The US Bureau of Labor Statistics (2015) compared wages in the manufacturing sector and found that the average Mexican manufacturing worker received less than 18 percent (US $6.36) of the hourly wage the average American manufacturing worker received in 2012 (US$35.67). The maquiladoras lined up along the Mexican side of this border exemplify how global businesses are taking advantage of these wage differentials. The operators of these manufacturing centers benefit from the lower labor costs in Mexico and the lenient enforcement of Mexico's otherwise stringent labor standards. The geographical proximity to the United States and Mexico's membership in the North American Free Trade Agreement provide easy and tariff-free access to the US market. Not only American businesses are seeking to benefit from the differences this border creates. A German business magazine recently promoted Mexico as an investment opportunity for medium-sized, often family-owned, German businesses because of the low wages, which have only increased "modestly" compared to China where annual wage increases have been larger. Furthermore, German entrepreneurs are being told that a business location in Mexico offers customs-free export to the potent consumer market in the USA. Therefore, it is "almost a must" for medium-sized German manufacturing suppliers to follow the lead of the larger German automobile manufacturers Volkswagen and Audi that have plant locations in Mexico (Markt und Mittelstand 2013).

The border has the effect that workers in developing countries like Mexico are denied access to the higher labor and wage standards in the countries of the global north, such as the United States. The border locks these workers into less favorable national labor markets (and often insufficient national healthcare and welfare systems, and lower living standards). Restrictions to cross-border mobility have created a "labor reserve" (Sassen 1988, 36) readily available for exploitation in the countries of the global south. Borders, in this way, enforce the international segmentation of labor into different wage and employment standards. "Unequal exchange" across borders subsequently ensures that a disproportionate share of the value the workers in the global south produce flows to the global north (Emmanuel 1972; Marx 1960 [1905–10]).

The second way in which borders control and discipline labor comes into force after workers cross borders. My previous research shows that the "international segmentation of labor" tends to persist even when workers manage to migrate from countries in the global south to countries in the global north (Bauder 2006). Although many skilled migrants obtain work permits or even immigration papers, crossing the border is often associated with a devaluation of their labor due to discrimination, social and cultural exclusion, and the non-recognition of their foreign credentials and work experience. Temporary foreign workers programs have a similar effect. The work of activist scholar Nandita Sharma (2006) illustrates how the Canadian temporary foreign workers program tends to deny workers the right to choose their employer, which, together with other program regulations and practices, effectively "bonds" these workers to their employers and prevents them from claiming rights that Canadian citizens would take for granted. The desolate conditions in the countries of origin – at the other side of the border – leave them little choice but to accept these conditions. Under one particular program, the Live-in Caregiver Program, many highly skilled women from the Philippines left their own children behind to work as domestic caregivers in Canada, often raising the children of Canadian families. They endure the deskilling of their labor (and sometimes physical and emotional abuse by their employers) because of the prospect of being allowed to stay permanently in Canada.

The situation is worse in the Gulf states, where a "sponsorship system," known as *kafala*, regulates labor standards for migrants in industries such as construction and occupations such as domestic work. The workers mentioned in the beginning of this chapter, who are building the stadiums and infrastructure for the 2022 FIFA World Cup in Qatar under slave-like conditions, exemplify the consequences of the *kafala* system. A recent report by Human Rights Watch (2014) documents the abuses suffered by domestic workers in the United Arab Emirates (UAE) under this system. Similar to some Canadian programs, this system effectively bonds the workers to a particular employer who possesses the authority to revoke sponsorship, resulting in the worker being deported. In addition, UAE's labor law excludes domestic work from basic labor and workplace protections, including limits to working hours and entitlement to overtime pay. While some employers may treat domestic workers well, the *kafala* system invites abuse and exploitation. Employers often confiscate the workers' passports, refuse to pay the full wages, and demand long working hours without adequate breaks. Workers have reported experiencing physical and verbal abuse, the withholding of food, and the denial of medical attention when they were sick or injured. Here is a case in the UAE reported by Human Rights Watch:

> Tahira S., an Indonesian worker, was subject to most of the indicators of forced labor. Her employer locked her inside the home and did not allow her out; shouted at, beat her, and broke a bone in her arm; confiscated her passport; made her work for 15 hours each day without rest periods or any days off and sleep on the floor with no blanket or mattress; gave her food only once a day and withheld it if her work was not deemed satisfactory; and promised to pay her only at the end of her contract but then paid her nothing. She told Human Rights Watch: "My boss started hitting me after two weeks of being there. Even though she hit me every day I wanted to wait for my salary. I thought if I waited three months I could get the money. She hit me with her fist to my chest. She scraped her finger nails to my neck, and slapped my face. I was bruised on my neck. She sometimes pulled out tufts of my hair."
>
> (Human Rights Watch 2014, 49)

Human Rights Watch estimates that at least 146,000 women from countries such as the Philippines, Indonesia, India, Bangladesh, Sri Lanka, Nepal, and Ethiopia are in the UAE as domestic migrant laborers. Throughout the Gulf region, workers who crossed borders experience that their labor is devalued and their human rights are trampled.

Workers who cross borders without the state's permission are sometimes in an even worse position. As "illegalized" migrants they often have no other choice but to work in the informal economy, where exploitation and abuse are rampant. Human Rights Watch reported about the worker John B., who was trafficked to the United States. Once he arrived there, he performed

> masonry and paving work in various states along the east coast. His bosses moved him and other workers from his home country between hotels every few weeks. John B. was prohibited from having relationships outside of work and was physically abused when he tried to do so. When he attempted to escape, his traffickers threatened to kill him and his family if he did not return to work.
>
> (Human Rights Watch 2010)

When he finally escaped this abusive work situation, he was apprehended by US immigration authorities, detained, and issued a deportation order. As an illegalized person, John B. lacked access to the protection by the state which citizens enjoy. Ruthless employers can take advantage of the vulnerable situation of workers such as John B.

Borders devalue labor and dehumanize workers because the global economy is dependent on a cheap and expendable work force (Cohen 1987, 135). The border ensures that this workforce either remains in the global south or, if it migrates to the global north, that it remains cheap and vulnerable. For many migrants, crossing the border signifies the moment when they lose their rights and their humanity.

Safe Haven

Another aspect of the border revolves around more positive emotions. The same migrants who experience the border as a mechanism of labor devaluation may also experience it as a gateway to safety and as a symbol of hope. Throughout human history people have crossed borders to escape the brutal and devastating consequences of war and persecution. After Adolf Hitler and his Nazi Party assumed power in Germany in 1933, many Jewish families and regime-critical individuals crossed borders to seek refuge from the brutal and anti-Semitic Nazi regime, migrating to the United States, Great Britain, and elsewhere. Many of these families and individuals had been members of the middle class in Germany, and relocation often meant the loss of their jobs, status, and social networks, and a life in poverty and social isolation. The philosophers Theodor W. Adorno and Hannah Arendt, whose scholarship I draw upon throughout this book, were refugees during this period. Together with other high-profile Jewish and regime-critical scholars and intellectuals, such as Albert Einstein and Berthold Brecht, they relocated from Germany to the USA to escape the Nazis. Their mutual friend and colleague Walter Benjamin was less fortunate. After crossing the French–Spanish border with the aim of continuing to travel to the USA, he learned that the Spanish government had prohibited border crossings. He feared that he would be returned to France and eventually extradited to Nazi Germany. Without hope, he reportedly killed himself with an overdose of morphine. The border as a gateway to safety had closed. Unfortunately, Benjamin did not know that the border had closed only temporarily.

More recently, the role of borders as a gateway to safety has been exemplified by China Keitetsi, who wrote a book about her horrifying experiences as a child soldier in Uganda. She describes how she crossed the border between Uganda and Kenya:

> I went straight to the border gate where I then successfully passed. When I had my foot on the Kenyan soil, I sat down and I do not remember who I thanked. My friends could only look, and shook their heads.
>
> (Keitetsi 2004, 246)

After crossing the border brought her to safety from immediate danger, she traveled by bus to South Africa, and then was resettled in Denmark by the United Nations. The refugees from Syria and elsewhere, described at the beginning of this chapter, also see the border as a gateway to safety and symbol of hope.

Even people who are dismissed as "economic migrants" and who cross the border well aware that their labor will be devalued and that they will be exploited, often see the border as a gateway to a better life. For example, the large income differential between Mexico and the USA – a manufacturing worker has more than five times the hourly wage in the USA than in Mexico – motivates many Latin Americans to cross the border to the United States, even if their entry is not authorized by American authorities. If these migrants receive, say, one-third of the hourly wages of their American counterparts and no legal protection or social benefits, migration may still be a considerable improvement of the circumstances. Similar calculations attract migrants to Europe and other rich destinations. These migrants endure the devaluation of their labor and the degradation of their humanity in exchange for the hope of a life with less material hardship. The border signifies this hope.

Marker of Distinction

Allow me to present a final example: the border can serve as a marker of distinction between different national identities and "cultural" practices. This aspect is experienced, for example, by travelers who leave the ordered beauty of San Diego in the USA and enter the colorful hustle and bustle of Tijuana in Mexico. Drawing on Hannah Arendt, John Williams (2006, 96) puts a positive spin on this aspect when he suggests that borders are "constitutive of a toleration of difference and diversity in human societies." Even in cases in which the border has physically disappeared and people are moving freely to and fro, this aspect of the border can be experienced. The Dutch–German border, for example, may be open to people but, as geographer Anke Strüver's (2005, 217) research shows, this border continues to divide "two nation states with different languages, norms, and habits."

Sandro Mezzadra and Brett Neilson (2013) refer to borders as a "method" that creates and enacts these differences to begin with. Thus, national identities and distinct cultural practices emerge after borders have been established. Along these lines, other researchers have shown how borders "actualize" (Shields 2006, 230), "institutionalize" (Eder 2006, 269), and "reify" (Anderson 1991) national identities and cultural differences. Even the mere visual contour of a country's border can function as a "logo" (Anderson 1991, 175) that evokes national pride and triggers patriotic emotions. The logo of Texas exemplifies this function, albeit at the state not the national scale (Figure 2.2).

However, this aspect of the border may be invisible for people crossing a section of the US–Canada border known as the Cascadian in Pacific Northwest of the North American continent. Geographer Matthew Sparke's (2005, 58) research proposes that the Cascadian can be seen as a region in which people share a common "state of mind" that evolved from the region's distinct ecology. To the residents of this region, the border as a marker of distinction may not exist. This example illustrates how the Cartesian border line that slices through the Cascadian along the 49th parallel (north) was drawn arbitrarily after the war of 1812 in a political effort to resolve boundary disputes between Britain and the United States. The common state of mind among the region's residents was apparently not a decision-making factor for drawing the border in this way.

The preceding examples show that border aspects are always context particular and incomplete. It is impossible to uncover an "essence" of the border. This impossibility, however, should not be interpreted as a failure to make sense of the phenomenon of the border. Quite the opposite: it allows us to think dialectically about borders and migration. The various aspects of the border capture partial truths about it, but each aspect is also limited in that it disregards other perspectives. A dialectic of the border concept is dynamic. New aspects emerge as migrants experience the border in new ways. Thus, a dialectic of the border is open-ended. Or, to use political geographer Gearóid ÓTuathail's (1999, 151) geographical metaphor, "new types of atlases" may need to be drawn as new border and migration practices emerge. In the next section, I examine this dialectical movement of borders and migration in greater detail.

Figure 2.2 Texas as logo
Source: Map by Harald Bauder

Border Dialectics

Contradiction is a central feature of dialectical thinking about borders and migration. The European colonizers of the late 19th century who drew arbitrary border lines across a map of Africa were in a very different situation than the desperate adolescent soldier who brought herself to safety by crossing a similar border, or the temporary foreign workers who cross a border to participate in a dehumanizing and exploitative labor program to feed their children. Despite their differences, these aspects of the border relate to each other, as in the contradiction between the free cross-border mobility of capital and the relative immobility of labor. This contradiction facilitates the accumulation of record profits for companies and their shareholders on the backs of workers who do not get their fair share of the value they create.

A Hegelian understanding of dialectics, however, has its limits in the way it applies to the concept of the border. According to such an understanding, contradiction is met by a solution – or "sublation" to use the philosopher's term – that mediates between the contradictory perspectives. Typically, this means that an entirely new vantage point offers a more comprehensive perspective that encompasses the preceding perspectives and thereby resolves the contradiction. I do not believe that a comprehensive meaning of the border concept can ever be achieved. The idealistic claim that the dialectic can lead us to a state where truth and thought conflate must be rejected in favor of positions that acknowledge the fragmented and political nature of knowledge (e.g. Foucault 1970, 1972). Critical theorists concur that it is futile to aspire to a point at which the dialectical process resolves into universal meaning (e.g. Horkheimer and Adorno 2004 [1947]). Instead, critical theory strives towards upholding oppositional, or negative, thinking and thereby affirms the continuation of the dialectical movement (Adorno 1963; Marcuse 1964). In the same way, we must recognize that the concept of the border is inherently unstable. A universal and fixed meaning of the border is neither attainable nor would it be desirable. Rather, we should accept from the outset that all aspects of the border are provisional.

An important idea in this book is that we can actively engage in the dialectical movement. The ability to notice multiple aspects of the border and migration is an important first step towards such critical engagement. The possibility of active engagement, however, raises important questions about our politics of engagement. In the complex "web of human relations," Hannah Arendt (1998 [1958], 183) warns us, any effort of proactive engagement will inevitably produce unintended consequences. Therefore, we must continuously *stay* engaged and critically reflect on the way we do.

The idealist Hegel does not offer much guidance on how this engagement ought to occur. Hegel (1970 [1820]; 1961 [1837]) believed that philosophers like himself are only passive bystanders who cannot anticipate the new or shape the dialectical movement. The famous epigraph that opened Part I of this book illustrates this belief. Karl Marx and Friedrich Engels responded to

Hegel's passive idealism in a manner that is more useful for my purpose. Following Ludwig Feuerbach (1986 [1841]), they famously sought to turn Hegel's dialectics from the head to the feet – or from "descending from heaven to earth" to "ascending from earth to heaven" (Marx and Engels 1953, 22, my translation). In this way, Marx and Engels emphasized that the dialectic has its roots in worldly circumstances rather than in the human mind. I applied this worldly grounded dialectic to the various border aspects I described earlier: each aspect derives meaning from the uses, experiences, and material practices related to borders. However, unlike Feuerbach – who, like Hegel, considered himself a passive observer of the dialectical process – Marx understood his scholarship as an educational activity that has the capacity to transform worldly conditions (Marx 1964 [1845]). In Marx's eyes, engagement in the dialectical process is a key responsibility of the critical scholar. Marx expressed this responsibility in his 11th thesis, which I presented as the second epigraph at the beginning of Part I of this book.

This break with Hegel's and Feuerbach's passive scholarship has presented a model for critical border and migration scholars, such as Étienne Balibar (2002, 2004). The way one sees a border is not a mere mechanical reflection of worldly circumstances and practices but also the product of our imagination. To use an earlier example: the colonizers of Africa assumed the cartographer's view, imagining the border as a line in Cartesian space, which then inspired them to draw the *actual* border between Namibia and Botswana as a straight line following the 20th and 21st meridians (east). The drawing of the actual US–Canadian border as a straight line through the Cascadian along the 49th parallel (north) followed a similar imagination. Although these lines disregarded the territories of ethnic or tribal communities and the shared "state of mind" among people living in a region, they created new realities and facts. Our imagination of the border also shapes worldly practices associated with it, including people's cross-border mobility.

Conclusion

There are several conclusions I draw from the above discussion. First, we must resist the urge to articulate fixed meanings of the border and migration. Such a pursuit would be futile. By affirming the ambiguity of the border concept, we acknowledge that the meanings of borders are always tied to particular uses, practices, and experiences. In this way, border aspects always remain connected to their particular contexts.

Second, we critical border and migration scholars, activists, politicians, and anyone else who engages the border dialectic can present fresh imaginaries of borders and migration, and these ideas may, in turn, affect material uses of borders and practices of migration. We can anticipate aspects of the border that challenge and potentially transform existing uses, practices, and experiences related to borders and migration. We can draw the atlases that make these aspects noticeable.

Third, the prospect of active engagement in the border dialectic raises the question of what kind of aspects of the border and migration would be worthwhile to present. Chris Perkins and Chris Rumford (2013, 274) remark in this context: "It is all well and good making a claim that this or that aspect of the border enables this or that form of action … The success of that claim lies with the actor's ability to make a case for its reasonableness." Elsewhere I have proposed a rather lofty vision of a democratic but non-state-centered border (Bauder 2011). In Chapter 5, I present a vision that may be considered more "reasonable" and takes the territorial state seriously. This vision appreciates the potential the state possesses to accommodate migrants who cross borders in the hope to reach safety, a better life, or both. Thereafter, in Chapters 6 and 7, however, I will drop all pretenses and dive into "Possibilia," where the territorial state as we know it today is no longer the political imagination framing borders, migration, and belonging.

References

1951 Refugee Convention, United Nations High Commissioner for Refugees. Accessed January 21, 2016. http://www.unhcr.org/pages/49da0e466.html.

Adorno, Theodor W. 1963. *Drei Studien zu Hegel* [Three Studies on Hegel]. Frankfurt am Main: Surkamp Verlag.

Alberts, Sheldon. 2006. "Canada–U.S. Border Seems to Be Missing." *CanWest News Service*, October 7. Accessed October 7, 2006. http://www.canada.com/story_print.html?id=17c5f136-b517-4aea-bf22-770c658be52b&sponsor=.

Anderson, Benedict. 1991. *Imagined Communities: Reflections on the Origin and Spread of Nationalism*. London: Verso.

Anderson, Malcolm. 1996. *Frontiers: Territory and State Formation in the Modern World*. Cambridge: Polity Press.

Arendt, Hannah. 1998 [1958]. *The Human Condition*. Chicago, IL: University of Chicago Press.

Balibar, Étienne. 1998. "The Borders of Europe." In *Cosmopolitics: Thinking and Feeling beyond the Nation*, edited by Pheng Cheah and Bruce Robbins, 216–233. Minneapolis, MN: University of Minnesota Press.

Balibar, Étienne. 2002. *Politics of the Other Scene*. London: Verso.

Balibar, Étienne. 2004. *We, the People of Europe: Reflections on Transnational Citizenship*. Princeton, NJ: Princeton University Press.

Bauder, Harald. 2006. *Labor Movement: How Migration Regulates Labor Markets*. New York: Oxford University Press.

Bauder, Harald. 2011. "Towards a Critical Geography of the Border: Engaging the Dialectic of Practice and Meaning." *Annals of the Association of American Geographers* 101(5): 1126–1139.

Brunet-Jailly, Emmanuel. 2005. "Theorizing Borders: An Interdisciplinary Perspective." *Geopolitics* 10: 633–649.

Bureau of LaborStatistics. 2015. "International Comparisons of Hourly Compensation Costs in Manufacturing, 2012." United States Department of Labor. Accessed December 15, 2015. http://www.bls.gov/fls/ichcc.htm.

Cohen, Robin. 1987. *The New Helots: Migrants in the International Division of Labour*. Aldershot: Avebury/Gower.

Dauvergne, Catherine. 2007. "Citizenship with a Vengeance." *Theoretical Inquiries in Law* 8(2): 489–506.

Dauvergne, Catherine. 2008. *Making People Illegal: What Globalization Means for Migration and Law*. New York: Cambridge University Press.

Eder, Klaus. 2006. "Europe's Borders: The Narrative Construction of the Boundaries of Europe." *European Journal of Social Theory* 9(2): 225–271.

Emmanuel, Arghiri. 1972. *Unequal Exchange: A Study of the Imperialism of Trade*. New York: Monthly Review Press.

Feldman, David. 2003. "Was the Nineteenth Century a Golden Age for Immigrants? The Changing Articulation of National, Local and Voluntary Controls." In *Migration Control in the North Atlantic World*, edited by Andreas Fahrmeir, Oliver Faron, and Patrick Weil, 167–177. New York: Berghahn Books.

Feuerbach, Ludwig. 1986 [1841]. *Das Wesen des Christentums* [The Essence of Christianity]. Ditzingen: Reclam.

Foucault, Michel. 1970. *The Order of Things: An Archaeology of the Human Sciences*. New York: Pantheon Books.

Foucault, Michel. 1972. *The Archaeology of Knowledge*. New York: Harper Colophon Books.

Geographer. 1972. *International Boundary Study No. 120: Angola–Namibia (South-West Africa) Boundary*. Washington, DC: Department of State, Office of the Geographer, Bureau of Intelligence and Research.

Haraway, Donna. 1991. *Simians, Cyborgs and Women: The Reinvention of Nature*. New York: Routledge.

Hegel, Georg W. F. 1961 [1837]. *Vorlesungen über die Philosophie der Geschichte* [Lectures about the Philosophy of History]. Stuttgart: Reclam.

Hegel, Georg W. F. 1970 [1820]. *Grundlinien der Philosophie des Rechts oder Naturrecht und Staatswissenschaft im Grundrisse* [Contours of the Philosophy of Right or Natural Right]. Stuttgart: Reclam.

Hegel, Georg W. F. 2005 [1807]. *Phänomenologie des Geistes* [Phenomenology of Spirit]. Paderborn: Voltmedia.

Horkheimer, Max and Theodor W. Adorno. 2004 [1947]. *Dialektik der Aufklärung: Philosophische Fragmente* [Dialectic of Enlightenment: Philosophical Fragments]. Frankfurt am Main: Fischer Verlag.

Human Rights Watch. 2010. "Victims of Trafficking Held in ICE Detention: Letter to the US Department of State on 2010 Trafficking in Persons Report." April 19. Accessed December 7, 2015. https://www.hrw.org/news/2010/04/19/us-victims-trafficking-held-ice-detention.

Human Rights Watch. 2014. "'I Already Bought You': Abuse and Exploitation of Female Migrant Domestic Workers in the United Arab Emirates." Report. Accessed December 7, 2015. https://www.hrw.org/sites/default/files/reports/uae1014_forUpload.pdf.

Johnson, Corey, Reece Jones, Anssi Paasi, Louise Amoore, Alison Mountz, Mark Salter, and Chris Rumford. 2011. "Interventions on Rethinking 'the Border' in Border Studies." *Political Geography* 30: 61–69.

Keitetsi, China. 2004. *Child Soldier*. London: Souvenir Press.

Marcuse, Herbert. 1964. *One-Dimensional Man: Studies in the Ideology of Advanced Industrial Society*. Boston, MA: Beacon Press.

Markt und Mittelstand. 2013. "Expansion nach Mexico: Was Mittelständler wissen müssen." March 7. Accessed December 16, 2015. http://www.marktundmittelstand.de/zukunftsmaerkte/expansion-nach-mexiko-was-mittelstaendler-wissen-muessen-1189651/.

Marx, Christoph. 2010. "Grenzen in Afrika als Last und Herausforderung." Heinrich Böll Stiftung. May 3. Accessed December 7, 2015. https://www.boell.de/de/navigation/afrika-grenzen-nationalstaat-afrika-kolonialismus-9109.html.

Marx, Karl. 1960. *Theorien über den Mehrwert*, Volume 3. Berlin: Dietz (originally written in 1861–3 and published in 1905–10).

Marx, Karl. 1964 [1845]. "Thesen über Feuerbach" [Theses on Feuerbach]. *Marx-Engels Werke*. Band 3. Berlin: Dietz Verlag. Accessed February 2, 2006. www.mlwerke.de.

Marx, Karl and Friedrich Engels. 1953. *Die deutsche Ideologie* [The German Ideology]. Berlin: Dietz Verlag.

Meier, Barry. 2015. "Labor Scrutiny for FIFA as a World Cup Rises in the Qatar Desert." *New York Times*. July 15. Accessed December 9, 2015. http://www.nytimes.com/2015/07/16/business/international/senate-fifa-inquiry-to-include-plight-of-construction-workers-in-qatar.html.

Mezzadra, Sandro and Brett Neilson. 2013. *Border as Method: or, the Multiplication of Labor*. Durham, NC: Duke University Press.

Neuman, Gerald L. 2003. "Qualitative Migration Controls in the Antebellum United States." In *Migration Control in the North Atlantic World*, edited by Andreas Fahrmeir, Oliver Faron, and Patrick Weil, 106–115. New York: Berghahn Books.

Newman, David and Anssi Paasi. 1998. "Fences and Neighbours in the Postmodern World: Boundary Narratives in Political Geography." *Progress in Human Geography* 22(2): 186–207.

Nicol, Heather N. and Julian Minghi. 2005. "The Continuing Relevance of Borders in Contemporary Context." *Geopolitics* 10: 680–687.

ÓTuathail, Gearóid. 1999. "Borderless Worlds? Problematising Discourses of Deterritorialisation." *Geopolitics* 4(2): 139–154.

Pattisson, Pete. 2013. "Revealed: Qatar's World Cup 'Slaves.'" *Guardian*, September 25. Accessed December 9, 2015. http://www.theguardian.com/world/2013/sep/25/revealed-qatars-world-cup-slaves.

Pérez-Peña, Richard. 2015. "Migrants' Attempts to Enter U.S. via Mexico Stoke Fears about Jihadists." *New York Times*. November 19. Accessed December 9, 2015. http://www.nytimes.com/2015/11/20/world/americas/migrants-attempts-to-enter-us-via-mexico-stoke-fears-about-jihadists.html.

Perkins, Chris and Chris Rumford. 2013. "The Politics of (Un)fixity and the Vernacularization of Borders." *Global Society* 27(3): 267–282.

Rose, Gillian. 1997. "Situating Knowledges: Positionality, Reflexivities and Other Tactics." *Progress in Human Geography* 21: 305–320.

Rumford, Chris. 2008. "Introduction: Citizen and Borderwork in Europe." *Space and Polity* 12(1): 1–12.

Sassen, Saskia. 1988. *The Mobility of Labor and Capital: A Study in International Investments and Labor Flows*. Cambridge: Cambridge University Press.

Sharma, Nandita. 2006. *Home Economics: Nationalism and the Making of "Migrant Workers" in Canada*. Toronto: University of Toronto Press.

Shields, Rob. 2006. "Boundary-Thinking in Theories of the Present: The Virtuality of Reflexive Modernization." *European Journal of Social Theory* 9(2): 223–237.

Smale, Alison and Kimberly Bradley. 2015. "Refugees across Europe Fear Repercussions from Paris Attacks." *New York Times*, November 18. Accessed December 8, 2015. http://www.nytimes.com/2015/11/19/world/europe/refugees-paris-attacks.html.

Sparke, Matthew. 2005. *In the Space of Theory: Postfoundational Geographies of the Nation-State*. Minneapolis, MN: University of Minnesota Press.

Strüver, Anke. 2005. "Bor(der)ing Stories: Spaces of Absence along the Dutch–German Border." In *B/ordering Space*, edited by Henk van Houtum, Olivier Kramsch, and Wolfgang Zierhofer, 207 221. Farnham: Ashgate.

Surk, Barbara and Rick Lyman. 2015. "Balkans Reel as Number of Migrants Hits Record." *New York Times.* October 27. Accessed December 9, 2015. http://www.nytimes.com/2015/10/28/world/europe/balkans-slovenia-reel-as-number-of-refugees-migrants-hits-record.html.

Torpey, John. 2000. *The Invention of the Passport: Surveillance, Citizenship and the State.* Cambridge: Cambridge University Press.

van Houtum, Henk, Olivier Kramsch, and Wolfgang Zierhofer, eds. 2005. *B/ordering Space.* Farnham: Ashgate.

Vaughan-Williams, Nick. 2008. "Borderwork beyond Inside/Outside? Frontex, the Citizen-Detective and the War on Terror." *Space and Polity* 12(1): 63–79.

Wastl-Walter, Doris. 2011. *The Ashgate Research Companion to Border Studies.* Farnham: Ashgate.

Williams, J. 2006. *The Ethics of Territorial Borders: Drawing Lines in the Shifting Sand.* Basingstoke: Palgrave Macmillan.

World Bank. 2015. World Development Indicators 2014. Accessed December 15, 2015. http://data.worldbank.org/data-catalog/world-development-indicators/wdi-2014.

Zollberg, Aristide R. 2003. "The Archaeology of 'Remote Control'." In *Migration Control in the North Atlantic World*, edited by Andreas Fahrmeir, Oliver Faron, and Patrick Weil, 195–222. New York: Berghahn Books.

3 Access Denied!

> Sooner or later, immigration controls will be abandoned as unworkable, too expensive in suffering and money, too incompatible with the ideals of freedom and justice, and impossible to maintain against pressures of globalization.
>
> Teresa Hayter (2001, 150)

The tragic deaths of thousands of migrants – ranging from the drownings of men, women, and children in the rough waters of oceans and seas, the deaths from dehydration in isolated desert regions, and the suffocation of travelers stowed away in shipping containers – illustrate the catastrophic human consequences of the border regimes that inhibit people from freely crossing international borders. Opening borders to all migrants would have prevented many of these deaths.

The case for open borders has been made from a remarkable array of ideological positions. Even people who consider themselves to be on opposite ends of the political spectrum can agree that borders should be open to everyone. In this chapter, I explore their arguments and the different paths they are taking to arrive at the same conclusion: that borders should be open. This exploration connects to the preceding chapter, in which I showed how the border assumes different meanings depending on an observer's vantage point. In the current chapter, I expand on the idea that borders and cross-border migration can be approached from different angles, narrowing my focus to arguments that support "open borders."

While exploring these arguments for open borders, we must not lose sight of the reasons why borders are not open. Today's borders maintain many of the political relations reminiscent of the world's colonial and imperial past. They disproportionately constrain the mobility of citizens of formerly colonized countries in the global south. In fact, some commentators suggest that current border practices reinforce a system of global apartheid (van Houtum 2010; Loyd et al. 2012). Similarly, migration and border restrictions reinforce economic inequalities. These restrictions tend to lock a vulnerable and exploitable labor reserve into the countries of the global south, maintaining an international segmentation of labor. Migrants who refuse to be deterred by this instrument of control and cross the border without state authorization often risk their lives, such as the migrants from Africa and the Middle East trying reach Europe, from Latin America attempting to enter the United States, or from Asia seeking refuge in Australia. If they manage to arrive at their desired destination, they are often illegalized and criminalized. And those migrants who are legally permitted to cross the border – such as the foreign workers in the Gulf states – often experience exploitation and abuse, while their labor contributes to the social and economic well-being of their employers.

Migration and border restrictions exist, according to economist John Isbister (1996, 57), because it is "in the interest of the privileged to protect their privileges." Activist Teresa Hayter (2001, 155) echoes this critique of migration and border controls: "The assumption of a moral right to impose suffering to preserve the privileges of a rich minority of course needs questioning." Since Isbister and Hayter made these observations around the turn of the millennium, borders have become considerably more brutal and deadly. The disproportionate benefits for the privileged, for whom borders tend to be open, on the backs of the underprivileged, for whom borders are closed, must be questioned more than ever. If borders were open to everyone, this instrument of denying people freedom and their humanity, exploiting their labor, and refusing them safety and security would lose much of its force.

Calls for Open Borders

The European Union's Schengen Area is an example of open borders being implemented because it was seen as politically feasible and economically advantageous to grant freedom of migration to people within this territory. In some other cases, friendly countries issue visa and work permits to each other's citizens on a relatively unrestrictive basis, and cross-border travel entails little more than a brief stop at the border crossing or point of entry and a short conversation with a border or immigration official. As a general demand, however, calls for open borders are typically dismissed as unrealistic and pushed to the margins of mainstream political debate and activism. Yet, in academic circles, the open-borders idea has received considerable attention (ACME 2003; Johnson 2003; Pécoud and de Guchteneire 2007). It has also been a topic of vivid discussion in internet-based forums, such as the website Open Borders (http://openborders.info/) and its associated Twitter (@openbordersinfo) and Facebook feeds. These calls for open borders do not necessarily suggest that there should be no border checks at all. Criminals (who would be prosecuted for the crimes they committed whether they migrate or not) could still be apprehended at the border. Rather, an open-borders scenario would grant all persons the same general freedom to migrate across international borders.

Despite the considerable support for open borders, there is no cohesive position on *why* borders should be open. Rather, the calls for open borders follow diverse and fragmented lines of reasoning. In introducing the topic of open borders, political philosopher Brian Barry (1992, 3–4) already marveled a quarter century ago that "it is not often that one is able to see how a number of very different approaches arrive at conclusions about a common set of problems." Below I review some of the positions from which calls for open borders have been made.

Liberal Positions

An important argument in support of open borders is that mobility constraints violate the core philosophical principles of liberalism, and thus nation states that claim to embrace these liberal principles cannot justify constraining cross-border migration. One of the most fundamental liberal principles is the moral equality of all human beings. This principle, however, appears to be at odds with borders that are open to some persons but closed to others. Selective migration policies and border regulations are especially worrisome if they are based on inherited privilege. The political scientist Joseph Carens (1987) has been an early advocate for open borders. His path-breaking work suggests that treating citizenship – and the associated right to enter and remain in a national territory – as a birthright is akin to feudal privilege, which liberalism strongly opposes. Carens draws on a number of liberal political theorists to make his point. Following the philosopher Robert Nozick, he blasts the idea that citizens somehow possess a collective birthright to the property of their national territory, and that only citizens have the right to cross international state boundaries or selectively deny entry to non-citizens. Equally worrying as birth privilege is that people are granted the privilege to cross international borders based on arbitrary criteria, such as possessing certain skills or money. Building on philosopher John Rawls' (1971) work on a free and rational society – but relaxing Rawls' assumption of a bordered political system – Carens concludes that in a global community of humanity, freedom of migration is a basic liberty.

Migration policies and border regulations are, by definition, exclusionary and treat human beings unequally. They routinely and openly violate universal ideas of equality. Liberal thinkers therefore have difficulties reconciling these policies and regulations with their guiding principles. In the words of philosopher Phillip Cole (2000, 3), "there is a serious gap between the legal and social practices of immigration and naturalization in those states that describe themselves as liberal democracies, and the fundamental commitments of a recognizable liberal political theory."

In his original work on open borders, Carens (1987) added a utilitarian argument in defense of free human mobility. When migration benefits both migrants and the receiving society, then migration is associated with a significant utility to both movers and non-movers. In many other cases, however, the citizens of a nation state may receive benefits by denying – rather than granting – migrants entry at their borders. Nevertheless, these benefits to citizens will most likely be smaller than the disadvantages experienced by potential migrants when they are denied crossing the border. In this case, closing the border to the migrants diminishes the aggregate utility for citizens and migrants taken together. Conversely, open borders would maximize the overall collective utility of all the persons involved in and affected by the migration process. Therefore, Carens concludes, borders should be open.

Other liberal proponents of open borders have presented a rights-based argument that implies that freedom of migration is a basic human right (Torresi 2010). Legal scholar Satvinder Juss supports this claim by illustrating that free movement has been "the historical norm in human society" (Juss 2004, 292). He argues that neither Biblical nor Roman nor medieval European legal practices restricted migration to the degree that today's nation states do. He further shows that the classical European publicists of the 16th to 18th centuries, such as Hugo Grotius in the Netherlands, Francisco De Vitoria in Spain, Samuel von Pufendorf in Germany, Emer de Vattel in Switzerland, or William Blackstone in England, spoke out against the emerging sovereign states' attempts to exclude aliens (Juss 2004, 297–302). Even the Prussian police regulation of 1932, on the eve of the Third Reich, permitted foreigners to stay on state territory as long as they "observe the laws and administrative regulations that apply on this territory" (Scherr 2015, 71, my translation).

Yet another liberal argument in defense of open borders takes an angle of applied ethics. According to this angle, border controls are "a prima facia rights violation" (Huemer 2010, 431) because they inflict harm by forcibly interfering with migrants' interests, including their legitimate pursuit to improve their lives or escape war, persecution, or poverty. From this point of view, the control of migration and borders "is very difficult to defend ethically because it is an institutional violation of the right to life and liberty" (Scarpellino 2007, 346). In certain circumstances the state may indeed interfere with the migration process – for example, if migration is an existential threat to the state. However, such circumstances are the exception to the default (i.e. *prima facia*) position of open borders; the burden of proof to make such an exception rests with the state (Ackerman 1980).

Proponents of liberalism have also presented counter arguments against open borders. However, these counter arguments are either based on weak evidence or they are inconsistent within the logic of liberalism. For example, one of these counter arguments suggests that migration is an external threat to the existence of the nation state. Following Thomas Hobbes (1969 [1651]), the state experiencing such a threat has the right to act in its own defense and restrict cross-border migration. Apart from displaying "moral partiality" (Cole 2000, 87) by privileging the principle of nationality over the principle of humanity, this argument against open borders can easily be challenged because migration rarely constitutes a threat to the very existence of the nation state. Granted, it may be a burden on a state's coffers or change the ethnic composition of its population, but it rarely threatens the state to a degree that it could cease to exist.

The following numbers illustrate this point. A series of Gallup polls has estimated how many migrants we might expect if borders were open globally.

One poll suggests that about 630 million people – or 13 percent of the world's adult population – would consider permanently moving to another country. Of these, 138 million would consider the USA as a destination, 42 million the United Kingdom, and 37 million Canada (Clifton 2013). Certainly, only a portion of the people who declared in the survey that they have a *desire* to move would actually do so if political borders were open. In fact, another Gallup report observed that many "people only dream of migrating"; only about 48 million globally are actually making preparations to migrate within a year (Ray and Esipova 2012). Open borders are unlikely to result in an immediate redistribution of the global population that might threaten the very existence of the receiving countries. These countries may gradually change as a result of migration, but they would survive.

Open borders have effectively existed between Puerto Rico and the United States since 1904. However, open borders did not spark immediate mass migration. In the first decade of the 20th century, only about 2,000 emigrants left Puerto Rico. The numbers slowly increased with a peak of roughly 470,000 in the decade between 1950 and 1960. By 1970 just below 1.4 million Puerto Ricans resided in the United States (Caplan 2014). As this example shows, open borders did not result in immediate population redistribution. Rather, the arrival of Puerto Ricans increased gradually over the period of several decades.

Europe provides another example of what might happen under an open-borders scenario. Fear mongers predicted dire consequences just before the Schengen agreement – which opened borders between member states – took effect. One can debate whether the actual numbers of migrants exceeded or fell short of the predictions. What matters is that inter-European migration rarely registers on the political radar. According to the German government, roughly 1,149,000 EU citizens entered Germany in 2014 while 472,000 left the country, representing a net gain of 667,000 migrants (BAMF 2015). Although these are significant numbers, neither German politicians nor the German media have constructed this migration as a problem. The narrative changed only when refugees from outside of the Europe Union entered Germany. In the fall of 2015, the German government expected to receive more than 800,000 asylum seekers and refugees from Syria and other countries for the year (the actual number ended up being higher), which dominated the news and preoccupied political debate for months.

During this refugee "crisis," the historian Paul Nolte remarked that "there are no objective limits to our ability to accept" newcomers (Nutt 2015, my translation). There are only short- and long-term consequences to consider. If Germany, for example, accepted two million refugees over three years, taxes may rise to pay for the housing of the refugees or class sizes in schools may increase to accommodate the refugee children in the school system. However, neither the political order nor the existence of the German state is fundamentally threatened.

One could still argue that a country like Canada, with a total population of about 36 million, would experience a considerable burden if only a small portion of the 37 million people who desire to go there actually arrived under the open-borders scenario. Nevertheless, a high annual intake of migrants could be construed as an opportunity rather than an existential threat. Examples in history show that immigration coincided with economic growth and gains in the geopolitical influence of the receiving states, without destroying these states' prosperity or their commitment to liberal democracy. The large-scale immigration of the 19th and early 20th centuries in the USA, at a time when borders were relatively open, also corresponded with the country's rapid industrialization and emergence as a major economic power in the world (Vineberg 2015). The immigrants provided not only needed labor but also skills, creativity, and entrepreneurship. Prominent examples of immigrants whose ingenuity contributed to the economic rise of America include the inventor Alexander Graham Bell, who revolutionized communication with inventions such as the telephone; industrialist and philanthropist Andrew Carnegie, whose investments helped create the powerful US steel industry; and the tailor and entrepreneur Levi Strauss, whose blue jeans transformed not only the garment industry but also established American leadership in the way people around the world would dress. If borders were opened today, there could be changes to the ethnic composition of the US population and structure and size of the country's economy, but not "a complete civilizational collapse or a revolution" (Smith 2015). Since its founding in 1776, the USA has always adapted to a growing and changing population. The state itself was never under threat from these transformations. On the contrary: immigration propelled it to become the dominant geopolitical force in the world today.

Even countries that are economically less prosperous than the USA, with fewer means to accommodate large numbers of migrants, do not collapse under the massive arrival of refugees who are escaping war and persecution by crossing international borders. The Office of the United Nations High Commissioner for Refugees (UNHCR) estimates that the ongoing fighting in Syria and Afghanistan had forced more than 1.8 million people into Turkey and more than 1.5 million into Pakistan in mid-2015. The small country of Lebanon, which has a total population of less than 6 million, has taken in an estimated 1.2 million refugees. The UNHCR estimates that Lebanon is hosting 209 refugees per 1,000 inhabitants (UNHCR 2015a). Many remain in these countries simply because more affluent countries, with a greater capacity to take in refugees, have largely closed their borders to them. While resources may be stretched when countries take in large numbers of refugees, the state is not endangered by the sheer numbers.

Polls have also estimated how many people would consider moving temporarily to a different country. These numbers are higher than those who consider permanent migration: about one in four or 1.1 billion people globally would like to move to another country temporarily for work (Ray and Esipova 2012). It is precisely this type of migration that is most often enabled by countries in the global north, because temporary migrants tend to benefit their national economies by supplying needed labor without "burdening" the state with the responsibilities it has towards its permanent residents and citizens. The large temporary migrant workforce that can be expected from open borders may strengthen rather than threaten the Hobbesian state.

The fact that liberal positions can be used to argue for and against open borders has created a "liberal paradox" (Basik 2013; Verlinden 2010). For example, there is a contradiction between the view that equal human beings possess the freedom of migration and the view that this freedom is a threat to the liberal state. In addition, the work of liberal political theorist Michael Walzer (1983) has been used to argue that national communities have the right to determine their own identity and membership by denying entry to migrants they do not want – and even to expel their own citizens for that reason (Hidalgo 2014). In this case, the liberal principles of human equality and freedom of migration conflict with the principle of community. Within the logic of liberalism – which tends to emphasize linear rather than dialectical thinking – these contradictions cannot been resolved at the theoretical level. Commentators have therefore suggested practical compromises, such as "fairly open borders" that permit some but not free cross-border mobility (Bader 1997).

Market-Economy Position

Much of the debate of open borders revolves around the economic impacts of freedom of migration. Economists who are participating in this debate often use the same argument for the mobility of labor that they apply to the free flow of capital and the trade of goods and services. Economic theory drawing on the work of David Ricardo suggests that the free geographical mobility of labor is "economically efficient for the world as a whole" (Gill 2009, 112). It permits regions and countries to specialize by allowing labor to migrate where it is needed most and can be used most effectively.

Conversely, migration and border controls distort the free labor market and therefore cause economic inefficiencies. By eliminating this source of distortion, open borders for labor have positive economic outcomes, such as increased total global incomes, reduced international wage differentials, and improved economic efficiency of national and global economies (Basik 2013). In this way, free international migration of labor serves the interests of individual workers, national economies, as well as global humanity as a whole.

This free-market argument for open borders is related to libertarian political philosophy, which proposes that preventing people from migrating freely is a form of violence that can only be justified if other people and their property are disproportionately affected by freedom of migration (Rothbard 1978). As long as the mobility of people across international borders is not immediately harming anyone or anyone's property, it should not be restricted. In the words of economist Jesús Huerta de Soto (1998, 192): "the ideal solution … would come from the total privatization of the resources which are today considered public, and the disappearance of state intervention at all levels in the area of emigration and immigration." Under such a scenario, the only condition of migration is that "the immigrant moves to a piece of private property whose owner is willing to take him [sic] in" (Block 1998, 173). This libertarian position towards property connects to the foundation of free-market capitalism: "Like tariffs and exchange controls, migration barriers of whatever type are egregious violations of laissez-faire capitalism" (Block 1998, 168).

Not only academics, but also politicians embrace versions of the free-market position towards cross-border labor mobility. None others than Ronald Reagan and George H. W. Bush supported such a position when they spoke at a Primary debate sponsored by the League of Women Voters in Houston, Texas, on April 4, 1980. When they were asked about their views on "illegal" migration, Bush said: "we have made illegal some kinds of labor that I'd like to see legal." Reagan followed up:

> Rather than putting up a fence [between the USA and Mexico] why don't we … make it possible for [Mexicans] to come here legally with a work permit, and then, while they are working and earning here, they pay taxes here, and when they wanna go back they can go back, and … [let's] open the border both ways.
>
> (Reagan and Bush 1980)

Open borders for the production factor labor is a common political position among free-market advocates.

Proponents of open borders who follow this market-economy position are not ignorant of the impacts of free labor mobility on a country's non-migrant population. In fact, they emphasize the positive nature of these impacts. Non-migrant property owners in migrant-receiving countries, for example, would benefit from increases in property values due to the higher demand for housing. In addition, migrants tend to contribute to a country's tax base and welfare system but, as non-citizens, they may never be eligible to receive corresponding benefits (Moore 1991; Riley 2008). Even in countries with strong welfare systems, open borders may not challenge these systems if welfare benefits and service provisions are conditional on having made prior contributions in the form of tax payments, insurance fees, or membership dues.

Political-Economy Position

Proponents of the political-economy position tend to be critical of market capitalism. Some of them are also suspicious of liberalism, including the liberal ideas of universal freedoms and equality. They see these ideas as manifestations of an ideology that affirms capitalism. They have instead focused on the economic exploitation of workers and the political oppression of people. According to this perspective, migrants experience social injustice because they are exploited and oppressed, and not necessarily because their inherent rights and equality are violated.

This argument relates to the aspect of the border that devalues labor, which I described in the preceding chapter. In this context, borders and migration restrictions have been tools of labor control and exploitation in the historical development of capitalism. The practice of restricting labor migration at international borders has served to separate a global labor force into competing national economies. This divide-and-conquer strategy aims to discipline the global labor force (Fahrmeir et al. 2003). More recently, border controls have reinforced the international segmentation of labor by locking vulnerable and exploitable workers into countries with low wage and labor standards, or by deskilling and criminalizing a considerable portion of the workers who are able to cross the border. In the meantime, "effectively, there are already open borders for highly-skilled workers," whom global capitalism favors (Castles 2003). While international borders tend to be permeable to privileged business and professional elites, border restrictions disproportionately target disadvantaged workers and facilitate their exploitation. Open borders would eliminate this mechanism of dividing the global labor force and rendering disadvantaged workers vulnerable and exploitable.

Proponents of the political-economy position also draw attention to the way global capitalism has uprooted and displaced people, causing these people to become migrants to begin with: "Today's so-called 'immigration problems' constitute only the tip of the iceberg of the enormous global chaos being created by ruthless forces of capital excess" (Darder 2007, 377–8). The liberalization of trade and the global advancement of capitalism have dispossessed and displaced large portions of the population of less-industrialized countries and thus compel the migration of people who would have preferred to stay. Karl Marx was already keenly aware of the displacement of traditional economies by the expansion of industrial capitalism. More recently, the expansion of capitalism and the associated displacement of non-capitalist economic practices have accelerated at a rapid pace. The geographer David Harvey (2005, 160–1) presents the well-known example of the reform of a Mexican law that had protected Indigenous peoples' right to collectively own and use land since 1917. The reform, which was initiated in 1991 by the Salinas government to prepare Mexico for the North American Free Trade Agreement (NAFTA), led to the uprising of the Zapatista Army of National Liberation in the southern Mexican state of Chiapas. On January 1, 1994, the day on which NAFTA took effect, the Zapatista rebels seized San Cristóbal de las Casas and other towns in the Mexican state of Chiapas in resistance to NAFTA and the Mexican government's politics. The privatization of land and the liberalization of trade with the United States would displace many small agricultural producers from their land in rural Mexico, adding to the kind of population pressure that results in increased migration.

Foreign competition and World Trade Organization rules have had similar effects on the economies and populations of other countries of the global south. Open borders for capital and trade have caused displacement from rural lands and destroyed the livelihood of many people. Open borders for people may not solve the root problem of displacement and poverty, but it would alleviate some of the labor market and population pressures experienced as a result of globalization.

Open borders would have additional benefits: many migrants send money to their family members and the communities they leave behind. The World Bank (2015) estimates that migrants have sent remittances to developing countries in the order of US$436 billion in 2014. These remittances are a significant flow of capital to these countries. They enable families to buy consumer goods and consume services, and they help communities to invest in education, infrastructure, and development projects. With the increase of migration under the open-borders scenario, global remittances would rise correspondingly.

The open-borders scenario would also make it easier for migrants to return to their places of origin. After all, they would know that they can migrate again if the situation at their place of origin requires them to do so – closed borders would not lock them in. Increased return migration can have positive effects on the origin country, since most migrants acquired "human capital" in the form of work experience, skills, education, specialized knowledge, and additional language proficiencies while they were abroad. When they apply this human capital in their countries of origin, it unfolds positive effects on the economy of these countries.

One must be careful, however, not to blindly endorse this idea of "brain circulation." This way of thinking can also be used to justify the selective migration policies of rich countries, who cherry-pick the "best and brightest" workers from the global south. In the end, the damage resulting from "brain drain," which these migration policies inflict on countries of the global south, may outweigh the benefit of return migration. A Gallup poll that asked more than 400,000 people in 146 countries about their desire to migrate to other countries revealed that mostly educated and professionally employed workers are making preparations to emigrate, not the uneducated and unskilled or the unemployed or underemployed who would benefit most by acquiring jobs and additional skills while abroad (Ray and Esipova 2012). The consequence is a brain drain: countries in the global south invest their meager resources in educating and training a skilled workforce, only to lose these investments when the workers move abroad. In 2006, approximately 167 Kenyan-educated medical doctors worked in the USA and the United Kingdom, representing an estimated total loss of more than US$86 million (in 2006 figures) to Kenyan society. Kenya lost an additional estimated US$411 million (in 2006 figures) due to the migration of nurses to seven Organisation for Economic Co-operation and Development countries. Additional damage results from the loss of health services to Kenyan society, the exodus of medical mentors and professional role models, and the reduced tax revenue (Kirigia et al. 2006). Ironically, many highly educated migrants from the global south cannot even apply their skills, education, experience, and credentials in their destination countries, resulting in a colossal "brain waste" (Reitz 2001).

Open borders would provide a level playing field for all potential migrants, counteracting the disproportionate departure of the educated elite and giving less-educated workers the opportunity to migrate to acquire valuable experience, skills, and knowledge abroad. If these formerly less-educated workers returned to their countries of origin as skilled and experienced workers, then these countries could experience a significant gain in human capital. In sum, open borders would create considerable increases in remittances and potentially enhance the circulation of "brains" that were upgraded rather than deskilled through the migration process.

Other Positions

There seem to be no limit to the range of positions from which it is possible to argue for open borders. The following arguments appear less frequently in scholarly or political debate than the preceding ones. The lower frequency, however, does not necessarily mean that these arguments have less intellectual or practical merit.

An anti-racist position suggests that "immigration controls have their origin in racism" (Hayter 2001, 149). According to this position, the most deadly borders are preventing the free migration of racialized people. The drawing of borders to separate people based on racial markers has a history that is entangled in colonialism. Such racially motivated border-drawing practices can be illustrated by the European settlers of North America, who banished Indigenous peoples behind the borders of "Indian reservations," or the South African apartheid system, which segregated racialized "blacks" into homelands. Today, border controls continue to disproportionately prevent racialized migrants from the global south from entering or settling in the countries of the global north.

Islamophobia is a related discriminatory practice with racial undertones that affects border practices today. Henk van Houtum (2010), for example, describes a blacklist of 135 countries whose citizens require a visa to enter the European Union's Schengen Area. When analyzing this list, he observed that Muslim countries are overrepresented. In other cases, racialized people are allowed to cross borders mostly under temporary migration schemes (Sharma 2005, 2006). Apparently, they are needed as workers but not wanted as citizens. Borders have "thus become essential institutions" enforcing a global system of apartheid (Balibar 2004, 113). Open borders would alleviate this racist practice.

Advocates of a gender argument for open borders emphasize that cross-border mobility restrictions disproportionately target women, and that restrictive migration policies often fail to assist and accommodate women and their children who are fleeing from war, violence, or poverty (Preston 2003). Slightly less than half of all migrants and refugees in the world are women, although women tend to be overrepresented among migrants in the global north (UNHCR 2015b; United Nations 2013). If women are able to cross borders, they often experience a double whammy: as migrants they are vulnerable and as women in patriarchal societies they are denied equal opportunities and perform undervalued service and care work. Often race and gender intersect, producing a triple whammy for migrant women who belong to racialized groups with origins in the global south. These intersecting vulnerabilities affect the Guatemalan woman who works informally as a maid in a Californian home, the Filipina nanny who takes care of the children of an affluent family in Singapore, and the Romanian mother who provides care services for the elderly in Italy. The care work of these women is essential for societies and economies to function. Yet, while this work enables others to get rich and live in comfort, the migrant women receive few benefits and obtain only a small fraction of the value they produce.

A political argument suggests that open borders can be advantageous for pragmatic geopolitical reasons. Open borders, for example, would deter military aggression if a state that invades another state would be expected to absorb the flow of displaced persons resulting from the invasion. The political geographer Nick Gill (2009, 113) remarks:

> Certainly, we might have expected the Western-backed wars in Afghanistan and Iraq to have received weaker support in America and the United Kingdom if these countries had expected to accommodate the majority of displaced persons created in these conflicts. For aggressive states who share land borders with countries they are considering invading, the likely consequences of war may very well prompt them to reconsider military action in a climate of reduced or absent border controls.

Open borders would in many cases shift a large portion of the burden of military intervention back to the aggressor.

Finally, religious and faith-based positions have supported open borders. For example, the idea of free migration has been related to statements made by Pope John Paul II and passages of the New Testament (Tabarrok 2000). Pope Francis too has been a vocal advocate for migrants – especially illegalized migrants – in the USA and elsewhere. The Roman Catholic Church's position on immigration reform in the United States has been interpreted as coming "close to an 'open borders' policy" (McGough 2014). Open borders can also be linked to Islam. In particular, the concept of *ummah* can be interpreted as describing a "translocal" global Muslim community unconstrained by national borders (Mandaville 2001). This concept has been related to cosmopolitan ideology that "traces back to the Ottoman period, which fostered an 'internationalist' atmosphere with its open borders" (Morey and Yaqin 2011, 180).

Conclusion

The prospect of open borders "is not a panacea for all issues of justice and equity" (Murphy 2007, 53). To alleviate the suffering, unequal treatment, and oppression experienced by many migrants would require more comprehensive measures than open borders alone. A comprehensive solution to these problems would simultaneously need to target the structural roots of various injustices, including human inequality, unfair competition, labor exploitation, racism, sexism, and military aggression (Gurtov 1991). The call for open borders is merely a limited response to the particular injustices and inequities created by current border policies and practices. They will not abolish all instances of human inequality, exploitation, racism, sexism, or violence. In this way, the calls for open borders represent only a limited political project.

The preceding discussion shows that various arguments for open borders draw on very different underlying philosophical positions to arrive at a similar conclusion: borders should be open. Some of these positions are irreconcilable with each other. For example, to the Marxist supporter of a political-economy argument for open borders, liberalism is merely an ideology that serves to justify the exploitation of workers and the appropriation of wealth under the guise of equality, freedom, and liberty. Yet, Marxist political economists, liberal theorists, and free-market economists can all agree on open borders.

Other philosophical positions related to calls for open borders are more compatible with each other. For example, liberal and free-market positions for open borders both center on the individual as the fundamental unit possessing inherent freedoms, including freedom of migration. Similarly, the liberal argument resonates with the political-economy argument for open borders in that both advocate for mitigating global poverty by helping the poor to escape economic despair by migrating to greener pastures (Wilcox 2009). Anti-racism and gender arguments for open borders can be aligned with a liberal position that perceives racial and gender discrimination as a violation of human equality. The rejection of racism and sexism is also compatible with the market-economy argument, which construes discrimination based on race and gender as a market distortion, denying meritorious but racialized and gendered labor equal access to the job market. Anti-racist and gender position can also be combined with a political-economy argument that rejects the social constructions of the categories of race and gender as integral to capitalist accumulation and exploitation. The connections between the various open-borders positions and arguments illustrate the complexity of the open-borders debate. However, each of these positions and arguments is also partial and incomplete, and the contradictions between them challenge us to think dialectically about the prospect of freedom of migration.

The various arguments for open borders and their multiple and pluralistic philosophical underpinnings reflect the ambiguous nature of borders, which was the subject of Chapter 2. In the same way that the border concept embodies various meanings that are all valid in the context of particular practices, it would be inaccurate to say that one argument for open borders is correct and the other ones are wrong. Furthermore, the integration of the various different arguments for open borders into a cohesive framework is neither possible nor desirable. What is worth emphasizing, however, is that the idea of open borders and free movement has proponents from across the political and philosophical spectrum. It is not an idea that can easily be dismissed as ideologically biased. This comprehensive support, I think, makes the idea of open borders a powerful inspiration.

But are open borders a utopian idea? If one tries to imagine what the concrete consequences of open borders could be, then the resulting picture can be disturbingly dystopian. The free mobility of people across international borders may produce a free-market dystopia in which the welfare state – an important achievement of the modern age – may collapse under the burden of an increasing number of welfare claims. The alternative of letting migrants cross borders but excluding them from welfare entitlement and denying them rights and citizenship is equally disturbing. Open borders could also increase labor competition among a global workforce, pitting migrant and non-migrant workers against each other, forcing them to accept ever lower wages and labor standards. In this way, open borders could intensify a global race to the bottom among workers (Hiebert 2003; Samers 2003). We can already catch a glimpse of such a dystopian future in the form of "American citizens who want to sell or rent their property to the highest bidders [and] the American businesses that want to hire the lower cost workers" (Binswanger 2006). From a "cultural" perspective, open borders could encourage national communities to deny migrants access to citizenship to protect the integrity of their "national cultures" (Vasilev 2015). Furthermore, open borders may exasperate efforts to tackle global economic inequalities: the very poor in the global south may not be able to afford to migrate, while the departure of the able-bodied, affluent, skilled, and educated would cause a drain within already poor societies of their meager monetary and human resources. As a result, open borders could trigger a cycle that concentrates poverty in poor countries and slows or even reverses the development of these countries (Bader 2005). These unsettling prospects illustrate that calls for open borders are a double-edged sword, which must be wielded with great care. The pathways towards the "utopian" possibilities of freedom of migration will be the topic of the next chapter.

References

Ackerman, Bruce. 1980. *Social Justice and the Liberal State*. New Haven, CT: Yale University Press.

ACME. 2003. "Engagements: Borders and Immigration Critical Forum on Empire." *ACME* 2(2): 167–220. Accessed January 27, 2016. http://acme-journal.org/index.php/acme/issue/view/47.

Bader, Veit. 1997. "Fairly Open Borders." In *Citizenship and Exclusion*, edited by Veit Bader. Basingstoke: Macmillan, pp. 28–60.

Bader, Veit. 2005. "The Ethics of Immigration." *Constellations* 12(3): 331–361.

Balibar, Étienne. 2004. *We the People of Europe?* Princeton, NJ: Princeton University Press.

BAMF (Bundesamt für Migration und Flüchtlinge). 2015. "Das Bundesamt in Zahlen 2014 Asyl, Migration und Integration." July 27. Accessed January 27, 2016. http://www.

bamf.de/SharedDocs/Anlagen/DE/Publikationen/Broschueren/bundesamt-in-za hlen-2014.pdf.

Barry, Brian. 1992. "A Reader's Guide." In *Free Movement: Ethical Considerations in the Transnational Migration of People and of Money*, edited by Brian Barry and Robert E. Goodin, 3–5. New York: Harvester Wheatsheaf.

Basik, Nathan. 2013. "Open Minds on Open Borders." *Journal of International Migration and Integration* 14(3): 401–417.

Binswanger, Harry. 2006. "Open Immigration." *Immigration Daily*. Accessed January 27, 2016. http://www.ilw.com/articles/2006,0329-Binswanger.shtm.

Block, Walter. 1998. "A Libertarian Case for Free Immigration." *Journal of Libertarian Studies* 13(2): 167–186.

Caplan, Bryan. 2014. "The Swamping that Wasn't: The Diaspora Dynamics of the Puerto Rican Open Borders Experiment." *Library of Economics and Liberty*. March 27. Accessed January 18, 2016. http://econlog.econlib.org/archives/2014/03/the_swamping_th.html.

Carens, Joseph H. 1987. "Aliens and Citizens: The Case for Open Borders." *Review of Politics* 49: 251–273.

Castles, Stephen. 2003. "A Fair Migration Policy—without Open Borders. Open Democracy." Accessed January 27, 2016. http://www.opendemocracy.net/people-migrationeurope/article_1657.jsp.

Clifton, Jon. 2013. "More than 100 Million Worldwide Dream of a Life in the U.S.: More than 25% in Liberia, Sierra Leone, Dominican Republic want to move to the U.S." *Gallup World*, March 21. Accessed January 27, 2016. http://www.gallup.com/poll/161435/100-million-worldwide-dream-life.aspx.

Cole, Phillip. 2000. *Philosophies of Exclusion: Liberal Political Theory and Immigration*. Edinburgh: Edinburgh University Press.

Darder, Antonia. 2007. "Radicalizing the Immigrant Debate in the United States: A Call for Open Borders and Global Human Rights." *New Political Science* 29(3): 369–384.

Fahrmeir, Andreas, Oliver Faron, and Patrick Weil, eds. 2003. *Migration Control in the North Atlantic World*. New York: Berghahn Books.

Gill, Nick. 2009. "Whose 'No borders'? Achieving Border Liberalization for the Right Reasons." *Refuge* 26(2): 107–120.

Gurtov, Mel. 1991. "Open Borders: A Global-Humanist Approach to the Refugee Crisis." *World Development* 19(5): 485–496.

Harvey, David. 2005. *The New Imperialism*. New York: Oxford University Press.

Hayter, Teresa. 2001. "Open Borders: The Case against Immigration Controls." *Capital and Class* 25(3): 149–156.

Hidalgo, Javier. 2014. "Self-Determination, Immigration Restrictions, and the Problem of Compatriot Deportation." *Journal of International Political Theory* 10(3): 261–282.

Hiebert, Daniel. 2003. "A Borderless World: Dream or Nightmare?" *ACME* 2(2): 188–193.

Hobbes, Thomas. 1969 [1651]. *Leviathan*. Menston: Scholar Press.

Huemer, Michael. 2010. "Is There a Right to Immigrate?" *Social Theory and Practice* 36(3): 429–461.

Huerta de Soto, Jesús. 1998. "A Libertarian Theory of Free Immigration." *Journal of Libertarian Studies* 13(2): 187–197

Isbister, John. 1996. "Are Immigration Controls Ethical?" *Social Justice* 23(3): 54–67.

Johnson, Kevin. 2003. "Open Borders?" *UCLA Law Review* 51(1): 193–265.

Juss, Satvinder S. 2004. “Free Movement and the World Order.” *International Journal of Refugee Law* 16(3): 289–335.

Kirigia, Joses M., Akpa R. Gbary, Lenity K. Muthuri, Jennifer Nyoni, and Anthony Seddoh. 2006. “The Cost of Health Professionals’ Brain Drain in Kenya.” *BMC Health Services Research* 6: 89–99.

Loyd, Jenna M., Matthew Michelson, and Andrew Burrigde. 2012. “Introduction.” In *Beyond Walls and Cages: Prisons, Borders, and Global Crisis*, edited by Jenna M. Loyd, Matthew Michelson, and Andrew Burridge, 1–15. Athens, GA: University of Georgia Press.

Mandaville, Peter. 2001. *Transnational Muslim Politics: Reimagining the Umma*. London: Routledge.

McGough, Michael. 2014. “On Immigration, Catholic Bishops Preach Gospel of (Mostly) Open Borders.” *LA Times*, April 5. Accessed February 27, 2016. http://www.latimes.com/opinion/opinion-la/la-ol-bishops-immigration-catholic-20140404,0,5200350.story#axzz2zqSYeEu3.

Moore, Stephen. 1991. “Immigration Policy: Open Minds on Open Borders.” *Business and Society Review* 77: 36–40.

Morey, Peter and Amina Yaqin. 2011. *Framing Muslims: Stereotyping and Representation after 9/11*. Cambridge, MA: Harvard University Press.

Murphy, Brian. 2007. “Open Migration and the Politics of Fear.” *Development* 50(4): 50–55.

Nutt, Harry. 2015. “Willkommenskultur und Selbstüberschätzung.” *Berliner Zeitung*, October 2.

Pécoud, Antoine and Paul de Guchteneire, eds. 2007. *Migration without Borders: Essays on the Free Movement of People*. New York: Berghahn.

Preston, Valerie. 2003. “Gender, Inequality and Borders.” *ACME* 2(2): 183–187.

Rawls, John. 1971. *A Theory of Justice*. Cambridge, MA: Harvard University Press.

Ray, Julie and Neli Esipova. 2012. “More Adults Would Move for Temporary Work than Permanently: About 1.1 Billion Worldwide Would Move for Temporary Work.” *Gallup World*, March 9. Accessed January 27, 2016. http://www.gallup.com/poll/153182/adults-move-temporary-work-permanently.aspx.

Reagan, Ronald and George Bush. 1980. “Primary Debate, Houston, Texas, on 04 April, sponsored by the League of Women Voters.” Accessed January 18, 2016. http://www.gettyimages.ca/detail/video/primary-debate-sponsored-by-the-league-of-woman-voters-news-footage/139842485.

Reitz, Jeffrey. 2001. “Immigrant Skill Utilization in the Canadian Labour Market: Implication of Human Capital Research.” *Journal of International Migration and Integration* 2(3): 347–378.

Riley, Jason L. 2008. *Let Them In: The Case for Open Borders*. New York: Gotham Books.

Rothbard, Murray N. 1978. *For a New Liberty*. New York: Macmillan.

Samers, Michael. 2003. “Immigration and the Spectre of Hobbes: Some Comments for the Quixotic Dr. Bauder.” *ACME* 2(2): 210–217.

Scarpellino, Martha. 2007. “‘Corriendo’: Hard Boundaries, Human Rights and the Undocumented Immigrant.” *Geopolitics* 12: 330–349.

Scherr, Albert. 2015. “Abschiebungen: Verdeckungsversuche und Legitimationsprobleme eines Gewaltakts.” In *Kursbuch 183. Wohin Flüchten?*, edited by Armin Nassehi and Peter Felixberger, 60–74. Hamburg: Murmann.

Sharma, Nandita. 2005. "Anti-Trafficking Rhetoric and the Making of a Global Apartheid." *NWSA Journal* 17(3): 88–111.

Sharma, Nandita. 2006. *Home Economics: Nationalism and the Making of "Migrant" Workers in Canada*. Toronto: University of Toronto Press.

Smith, Nathan. 2015. "How Would a Billion Immigrants Change the American Polity?" Open Borders: The Case (Blog), August 14. Accessed January 27, 2016. http://openborders.info/blog/billion-immigrants-change-american-polity/.

Tabarrok, Alexander. 2000. "Economic and Moral Factors in Favor of Open Immigration." *Independent Review*, September 14. Accessed January 27, 2016. http://www.independent.org/issues/article.asp?id=486.

Torresi, Tiziana. 2010. "On Membership and Free Movement." In *Citizenship Acquisition and National Belonging: Migration, Membership and the Liberal Democratic State*, edited by Gideon Calder, Phillip Stoke, and Jonathan Seglow, 24–37. Basingstoke: Palgrave Macmillan.

UNHCR. 2015a. *Mid-Year Trends 2015*. Geneva: UNHCR. Accessed December 21, 2015. http://www.unhcr.org/56701b969.html.

UNHCR. 2015b. *UNHCR Statistical Yearbook 2014, 14th Edition*. Geneva: UNHCR. Accessed December 21, 2015. http://www.unhcr.org/566584fc9.html.

United Nations. 2013. *International Migration Report 2013*. New York: Department of Economic and Social Affairs, Population Division. Accessed December 21, 2015. http://esa.un.org/unmigration/documents/worldmigration/2013/Full_Document_final.pdf.

van Houtum, Henk. 2010. "Human Blacklisting: The Global Apartheid of the EU's External Border Regime." *Environment and Planning D: Society and Space* 28(6): 957–976.

Vasilev, George. 2015. "Open Borders and the Survival of National Cultures." In *Rethinking Border Control for a Globalizing World: A Preferred Future*, edited by Leanne Weber, 89–115. Florence, KY: Taylor and Francis.

Verlinden, An. 2010. "Free Movement? On the Liberal Impasse in Coping with the Immigration Dilemma." *Journal of International Political Theory* 6(1): 51–72.

Vineberg, Robert. 2015. "Two Centuries of Immigration to North America." In *Immigrant Experiences in North America: Understanding Settlement and Integration*, edited by Harald Bauder and John Shields, 34–59. Toronto: Canadian Scholar's Press.

Walzer, Michael. 1983. *Spheres of Justice: A Defense of Pluralism and Equality.* Oxford: Martin Robertson.

Wilcox, Shelly. 2009. "The Open Borders Debate on Immigration." *Philosophy Compass* 4(5): 813–821.

World Bank. 2015. "Topics in Development: Migration, Remittances, Diaspora and Development." Accessed January 27, 2016. http://go.worldbank.org/0IK1E5K7U0.

4 From Utopia to Possibilia

> A map of the world that does not include Utopia is not worth even glancing at, for it leaves out the one country at which Humanity is always landing. And when Humanity lands there, it looks out, and, seeing a better country, sets sail. Progress is the realisation of Utopias.
>
> Oscar Wilde (1891)

In the preceding chapters, it became clear that international migration is controlled not only at the physical border line but also at airports and transit hubs before migrants reach the border line, and at workplaces and in public spaces after migrants have crossed the border line. Therefore, freedom of migration is associated with more than simply crossing the physical border. It also relates to other aspects of the border, including the ability to participate as an equal member in society and the labor market.

A world of free human mobility is not a simple utopia. There are many ways in which freedom of mobility can be imagined. An open-borders scenario, for example, assumes that nation states with territorial borders continue to exist and that everybody can freely cross these borders. Free human migration is also possible under the so-called no border scenario, which, as the name suggests, entails that there are no borders at all. This scenario requires the radical transformation of not only existing political circumstances but also the core ideas according to which societies organize themselves. The open-borders imagination thus affirms the territorial nature of governance, while a no-border imagination eliminates nation states and their borders altogether. In this chapter, I bring the prospects of open borders and no border into dialog with each other.

A Note on Utopia

The term *utopia* is a combination of the Greek words *eutopia* (good place) and *outopia* (no place). Thomas More (1997 [1516]) coined the term "Utopia" to describe a fictional island somewhere on the edges of the Atlantic Ocean. In the storyline of his book, More relays his conversations with the traveler Raphael, who had lived on the Island Utopia (depicted in Figure 4.1) for

Figure 4.1 The Island of Utopia, cover illustration of the first edition of More's book, 1516
Source: Wikimedia Commons

several years. On this island, inhabitants had established a society based on the principles of reason, where social, political, judicial, and economic systems differed from those in More's contemporary Europe. The society of the island Utopia offered religious co-existence, shunned private property, provided a structured work day, but endorsed slavery. For More, Utopia served as a tool to critique contemporary society by projecting an image of a society that is different.

Since More coined the term, the far-away world of Utopia has served as a powerful figure to critique existing society. In the early 20th century, the writer H. G. Wells described a modern Utopia in which the freedom of migration is an explicit feature. He submits that "to the modern-minded man it can be no sort of Utopia worth desiring that does not give the utmost freedom of going to and fro. Free movement is to many people one of the greatest of life's privileges" (Wells 1959 [1905], 34). Wells therefore concludes that "the population of Utopia will be a migratory population beyond any earthly precedent, not simply a travelling population, but migratory" (Wells 1959 [1905], 45). In a world in which freedom of migration is assumed, borders will not be seen as a problem. This situation may explain why borders have rarely been problematized whenever utopia has been concretely articulated (Best 2003). Still, I am puzzled by this lack of attention to borders, given that in today's world borders are so fundamental to the production and reinforcement of inequality, injustice, and oppression.

Conversely, the imagination of a world with open borders or without borders is often labeled "utopian." Commentators typically use this label to dismiss the open-borders and no-border ideas outright, without seriously engaging with them. Political scientist John Casey, for example, observes that "advocacy of a universal open border policy is seen at best as a policy-irrelevant chimera and utopia" (2009, 15) and that "any discussion of open borders is dismissed as 'pie in the sky' utopias" (42). As an example of the dismissive public attitude towards open borders, Casey (2009, 53) cites the Canadian newspaper, the *Globe and Mail*, which had suggested that free cross-border labor mobility would be "a utopian madhouse, even crazier in concept than communism." By calling an open-borders world utopian, it is presented as an absurdity.

In politics, the concept of utopia has often been used in a polemical way. For example, in the 19th and 20th centuries, socialist and communist visions were labeled "utopian" in a dismissive and derogatory manner. Even socialist sympathizers, like Karl Marx (1982 [1848]) and Friedrich Engels (1971 [1880/1882]), opposed utopia as an idealistic and dogmatic concept. More recently, utopia has been associated with failed totalitarian regimes, including Stalinism and Nazism. As a discredited practice, utopian thinking has largely disappeared from mainstream political discussions. Instead, political debate today presents market capitalism and territorially organized nation states as the only imaginable possibility. Margaret Thatcher's infamous proclamation that "there is no alternative" has come to symbolize the apparent foolishness of critiquing the dominant economic and geopolitical order of the world.

Utopia is no longer used explicitly as a tool to radically rethink today's society. Instead, utopian thinking surfaces in the guise of scientific impartiality that reaffirms the existing economic and political order. The economists Friedrich Hayek and Milton Friedman, for example, had visions of a free-market utopia, in which capitalist enterprise is not distorted by political interference. Armed with this vision, political and economic actors like Thatcher were able to further entrench free-market capitalism in political and economic life, which facilitated the advancement of neoliberal capitalism as we know it today (Harvey 2005).

And yet, implicitly, utopian thinking continues to play an important part in envisioning a possible world beyond the existing one. Philosophy of history scholar Cosimo Quarta suggests that utopianism is part of human nature, distinguishing humans from other species; utopia is engrained in humanity's restless "search for new possibilities" (Quarta 1996, 159). Especially in today's political climate, in which it seems foolish to think about alternatives, it is important to explore utopian possibilities. As Marcus Hawel and Gregor Kritidis (2006, 8, my translation) observe: "Only when we conceptually cross exiting boundaries, will we be able to unleash the forces necessary for the material transgression of these boundaries." Or, in David Harvey's words: "without a vision of Utopia there is no way to define that port to which we might want to sail" (Harvey 2000, 189).

So, what kind of utopia should we embrace? Utopia is an ambiguous concept. While forward looking and forward thinking may be a part of human nature, there is no single way in which people have imagined utopia. Utopia typically serves a dual role: first, it critiques contemporary society. In this way, utopia *negates* existing conditions deemed problematic. More's (1997 [1516]) utopia served in this role as a critique of contemporary Europe, lamenting its relation to private property and problematizing other political circumstances and social practices. Second, utopia defines an alternative ideal-type world; it shows how people *should* live with each other. More's description of Utopia also exemplifies such a better world in tangible ways. The fact that More's book was illustrated by the picture of Utopia (shown in Figure 4.1) indicates that this alternative world was conceived as a concrete object. Some scholars agree that the value of utopia lies in describing a concrete alternative society. The philosopher Richard Rorty, for example, suggests that it is not enough to voice critique, but that critique should be followed by concrete alternative suggestions:

> My own view is that it is not much use pointing to the "internal contradictions" of social practice, or "deconstructing" it, unless one can come up with an alternative practice – unless one can at least sketch a utopia in which the concept or distinction would be obsolete.
>
> (Rorty 1991, 16)

The geographer David Harvey, too, wrote about a utopian dream of a post-revolution world in which hierarchical political order and state-controlled borders have been demolished. In this dream, all people enjoy mobility between regions and nations – or what Harvey envisions as *regionas* and *nationas* – with the only limitation being that an electronic bulletin board manages the comings and goings of people between regions to balance skill levels and prevent regional economies from collapsing due to a massive brain drain (Harvey 2000, 257–81).

One must be mindful, however, of the consequences of constructing utopias. The critical theorist Theodor Adorno argues that utopia should *not* be articulated in concrete terms, as in "it will be like this and this [*so und so wird es sein*]" (Adorno and Bloch 2014). Concrete articulations of utopia will always need to use concepts and ideas that people are already familiar with, and they always rely on existing language and ways of thinking. These concrete utopias thus tend to reproduce the ideologies that already exist in our heads (Mannheim 1952 [1929]). Moreover, defining utopia as a concrete alternative, and acting to achieve it, stifles the free unfolding of other possible futures – especially a future that we yet lack the language and concepts to describe, let alone understand. The stone-age travelers to whom the earth appeared to be a flat disk could probably not comprehend that they could never reach an edge from where the earth drops off. Neither would a king who lived in medieval Europe have been able to imagine a modern democratic nation state, because the concept of modern democracy and the practice of organizing polities territorially did not yet exist. In the same way, we cannot describe the future with the terms and concepts available to us today. Theorists like Adorno (1966) therefore suggest that critique should remain at the stage of negation, which points out what should *not* be without defining how things should be different. Any attempt to translate critique into a concrete alternative is an ideological exercise. Only as negation will critique keep all alternative possibilities open.

Negation and Possibility

Calls for open borders and no border are first and foremost critiques of existing border regulations and of bordering practices that distinguish between people based on their place of birth, citizenship, ancestry, race, or wealth. The various arguments for open borders, which I reviewed in the previous chapter, call for an end to migration restrictions without developing alternative models of migration or governance. In other words, the calls for open borders *negate* the contemporary condition of closed and controlled borders. Because these calls do not present alternative worlds, they are not utopias in the conventional sense. They do not define the particular circumstances of how people should live together or how societies should be governed to achieve freedom of migration. Rather, they simply critique and reject the present limitations to the freedom of migration.

Representing the liberal position for open borders, the political scientist Joseph Carens remarks that his path-breaking critique of border controls was not intended for implementation in any concrete way. In an exchange with the economist John Isbister, he writes:

> The open borders argument is not really intended as a concrete recommendation for current policies or one in a foreseeable future. It is not intended as advice to presidents and prime ministers or to administrators and legislators. Rather, it serves a heuristic function, revealing to us

> something about the specific character of the moral flaws of the world in which we live, the institutions we inhabit, and the social situation of those who dwell in rich industrial states.
>
> (Carens 2000, 643)

Carens (2000, 637) further writes: "I imagine (or at least hope) that in a century or two people will look back upon our world with bafflement or shock" at the injustices that border controls inflict in today's world. Carens is voicing an internal critique of liberalism and refrains – at least in this particular publication – from speculating what a concrete open-borders future would look like. In this way, the open-borders argument remains a negation and leaves open the manner in which an open-borders future could unfold.

In contrast to the open-borders perspective, a no-border perspective follows a tradition of feminist, anti-racist, and anti-colonial scholarship and activism. It opposes borders altogether – and with them the territorial nation state that these borders define. No-border advocates see nation states and their borders as sources of oppression (Alldred 2003). National borders have created the category "migrant" to begin with: a migrant is someone who crosses a border. Without borders there would be no such labels as the "good" migrant who is hard working and docile, or the "bad" cue-jumping refugee, or the "economic migrant" who steals our hard-earned wealth. The no-border position rejects these labels created and imposed by borders. Activist scholars Bridget Anderson, Nandita Sharma, and Cynthia Wright (2009, 6) explain that "any study of national borders needs to start with the recognition that they are thoroughly ideological." As ideologies, national borders justify the practices of population control, labor exploitation, and national distinction that I have described in Chapter 2.

Open border and no border calls identify border controls and restrictions on the freedom of migration as one of the greatest and most deadly problems of our day. These controls and restrictions enable the unequal treatment of persons who are otherwise equal; they distort free markets; they facilitate labor exploitation; and they enforce oppression based on racial markers and gender. In this way, calls for open borders and no border negate the contemporary condition of closed and controlled borders and the unfreedom, inequality, social injustice, and oppression that border practices create. This rejection of current border practices and borders, however, is not necessarily tied to concrete blueprints of alternative worlds.

As pure negation, however, open borders and no-border scenarios say nothing about the conditions under which unconstrained human migration ought to occur. They do not convey how migration should be regulated, how sovereignty should be exercised, how labor markets should be managed, or how people should become members of territorial communities. As a pure negation, the "dream" of freedom of migration remains intangible.

To illustrate how utopian thinking is still relevant, I draw on the philosopher Ernst Bloch (1985 [1959]), who theorized "the possible" not as a single condition to which one aspires, but rather as a multidimensional category that involves existing and "not-yet" existing circumstances. In particular, Bloch's work enables me to explore different "layers" of possibility. One layer of interest is the "fact-like object-based possible" (*sachhaft-objektgemäß Mögliche*), which, for the purpose of clarity and readability, I call the "contingently possible." It refers to what is possible when certain conditions are met. This possibility requires an "inner" capacity to enact the possibility as well as the "outer" conditions under which this possibility can occur. Bloch used the following example to illustrate what he meant: a blooming flower has the inner capacity to ripen into a fruit but only under the outer condition of suitable weather and climate. Another example that resonates more with this book's topic is a society that has the inner capacity to provide freedom of migration, equality, and social justice under the outer condition of the territorial nation state. Similar to weather and climate, the territorial nation state is a condition that currently exists and that is therefore imaginable. However, unlike in biology, where the ripening of a flower into a fruit is a necessary and predictable outcome once all the required conditions are met, in the human world achieving the possible is a political activity; it requires creativity to mediate between inner capacity and outer material conditions. In other words, achieving the contingently possible is a dialectical process that involves human engagement.

Bloch distinguishes the layer of the contingently possible from what he calls the "objectively-real possible" (*objektiv-real Mögliche*). The term "real," in this context, does not refer to the actual world in which we live, but rather a possible world that is not reducible to particular aspects, such as a specific political system. I will call this layer "possibilia." The reason I decided to use possibilia is that the term linguistically connects utopia with possibility. This term is, to the best of my knowledge, usually not used in this way; it is also not applied to the contemporary debate of migration and borders. In philosophy this term sometimes refers to "objects of unrealized assemblages, of false but coherent scientific theories, or unfulfilled plans" (Voltolini 1994, 75) that can be rejected as things that are false or do not exist (Nute 1998, Marcus 1975–6). I use possibilia in a way that resembles more closely Bloch's ideas. Possibilia, to me, encompasses the totality of possible social and spatial relations. It differs from the contingently possible in that it assumes that political, social, and economic circumstances will be different from what they are today and that people will think differently about the world than they do now. Possibilia thus refers to a world that arises under not-yet-existing conditions and the not-yet-discovered ways of imagining this world (Bloch 1985 [1959], 274–5).

Let me return to the preceding example to illustrate the difference between the two layers of possibility. A seed already contains the genetic blueprint of an organism that produces a fruit under the right conditions. Similarly, the model of the territorial nation state harbors the capacity to grant certain degrees of freedom, equality, and justice to its citizens under certain circumstances, such as a liberal constitution, democratic voting, possibilities of political engagement, an uncorrupt bureaucracy, and a fair and just legal system. In contrast, possibilia projects an *open* future that does not rely on an existing blueprint. Rather, possibilia is based on conditions and practices that do not yet exist and that we cannot yet imagine with today's concepts and ways of making sense of the world.

Open Borders as Contingent Possibility

Many readers will probably find it hard to imagine a world in which territorial statehood is not the dominant political organizing principle. In this section, I therefore explore the contingent possibility of a territorial state with open borders. In other words, I do not question the existence of borders but rather assume that they are open. This also means that I refrain, for now, from anticipating a not-yet-existing political order that does not have territorial nation states or ideas of community and belonging that differ from the way we think of them today. It is not unreasonable to assume that territorial statehood remains the dominant political organizing principle, and that formal citizenship continues to define who is formally a member in this territorial state. In fact, any diversion from this assumption would typically be dismissed as unrealistic. Of course, that nations and states possess a territory is not a "natural" thing in the way that a Siberian tiger or certain species of insects claim territories. Rather, the sovereign territorial nation state is the outcome of a historical process that lasted centuries and gradually fused state and nation within a territorial container (Sassen 2006). This territorial nation state is now so deeply entrenched in our geopolitical imagination that it seems impossible to imagine a world order without it. Even globalization – which supposedly challenges the nation state – has been organized by territorial nation states (Paasi 2009).

Recall my earlier example explaining the contingently possible: we can think of the territorial nation state and formal citizenship as the "outer" frame, within which the "inner" capacity of freedom, equality, and justice rests. Currently, however, border controls and selective access to formal citizenship are major barriers to achieving this potential. The question we then need to ask is: How should the idea of open borders be implemented to enable nation states to achieve their inner capacity?

Concrete attempts to articulate an open-borders scenario do exist. Austro-Marxist Otto Bauer presented a socialist open-borders vision. As a Marxist, Bauer rejected utopia and instead stressed that his vision is not a "phantasy" but based on scientific reason and "sober assessment" (Bauer 1907, 521). According to his vision, cross-border migration follows a "conscious regulation of migrations … [that] will attract immigrants where the increased number of workers multiplies the productivity of labor" (Bauer 1907, 515, my translation). The state will no longer control labor flows in its own interests, and national communities will be severed from their current state territory and be mobile across state borders. In Bauer's vision, the principle of nationality is the rational organizing principle of a socialist society. This principle will "sweep away all traditional ideologies as soon as the dam of capitalism is broken" (Bauer 1907, 511, my translation). Bauer (1907, 520) further explains that in this socialist world, members of a nation will no longer migrate as individuals. Rather, they will migrate as a corporate-legal entity (*öffentlich-rechtliche Körperschaft*) which ensures that the migrant's cultural, social, and economic rights are protected.

We must remember that Bauer envisioned his future socialist world from the vantage point of the Austro-Hungarian Empire, preceding the catastrophes of WWI and WWII in which nationalism wreaked havoc on Europe and other parts of the world. In his socialist vision, Bauer sees the dominance of the principle of nationality as a historical necessity. As a consequence of this principle, nations possess the freedom of migration between state territories. The overarching – and I think problematic – principle of nationality excludes other possibilities of political and social organization and cross-border mobility.

Another concrete open-borders vision follows the free-market principle. More than a quarter century ago, the sociologist Ruth Levitas (1990, 186–7) observed that the "neo-liberal New Right" envisions a utopia of a society in which individuals freely compete with each other in the labor market without interference by the state. Under the conditions of open borders, the market principle can freely unfold and maximize the utility for workers, employers, and society as a whole. Similar to Bauer's vision of a future socialist world, which favored the principle of nationality, this free-market utopia does not permit any principles of regulation other than the market and thereby denies alternative possibilities of open-border worlds. Rather, Bauer's socialist utopia and the free-market utopia reproduce highly problematic existing ideologies.

Of course, as a contingent possibility, the concrete imagination of open borders relies on existing concepts and ways of thinking – in particular, the concept of the territorial nation state and certain organizing principles. Even if we assumed that borders are open and that entry into a territorial state is therefore free to everyone, then an important follow-up question arises: do states continue to exclude migrants because they deny access to rights and citizenship? Under current conditions, this may be the case. If states say: "you are permitted to cross the border but you will not be protected by our laws or permitted to participate in our welfare system or in our political life," then open borders will effectively establish a cast society in which migrants are subservient to citizens. Even Michael Walzer (1983, 52–61) – who otherwise argued against freedom of migration – concedes that foreign residents must be included in the political decision-making process; anything else would amount to tyranny. However, migration policies and border practices today are often designed to establish precisely such differences between migrants and citizens. Guest worker and temporary foreign worker programs, for example, deny migrants equal rights. In addition, laws that formally deny entry may in practice not prevent cross-border migration but serve to illegalize migrants after they crossed the border. In both cases, the result is a caste-like distinction between migrant and citizen.

Nevertheless, one could argue that today's states are at least heading in the right direction. After all, most states observe international law that grants migrants a set of basic rights even when they do not possess formal citizenship of the country in which they reside. In practice, however, migrants are in vulnerable positions that make it difficult to claim these rights. Illegalized migrants, for example, will not call the police to report a crime or sue an employer to receive their unpaid wages if they risk deportation. One could still argue that once foreigners gain *legal* access to a country they can acquire rights and entitlements. The sociologist Yasemin Nuhoğlu Soysal (1994, 12) found in her study of migrants in Europe that formal citizenship in a nation state "is no longer the main determinant of individual rights and privileges." Rather, personhood provides access to rights. Her research shows how non-citizen migrants acquire social, economic, and political rights through residency, making contributions to national social security and employment insurance systems, and participating in other ways in the national community. She used the case of Turks in Germany to illustrate her point. However, in this case, migrants were first carefully selected through a guest workers program and then had to undergo a series of probationary and stepwise residency and employment hurdles before they could accrue additional rights. In the USA, migrants also tend to receive rights gradually, in the form of an "incremental process from less to more" (Bosniak 2007, 291). During this probationary period migrants remain vulnerable.

Citizenship, too, is a matter of state discretion. Migrants must meet certain state-defined eligibility criteria for naturalization. For example, in the USA foreign permanent residents must wait for a five-year period before they are eligible for naturalization, and even then, citizenship can still be denied to them. The applicants must demonstrate "good moral character," which the US Department of Homeland Security (2015) instructs officers to assess based on whether the applicants have a criminal history and have complied with the conditions of probation, whether they abide by the law and are involved in the community, as well as factors such as education, employment history, family ties, and background. Temporary foreign residents and non-status residents may not be eligible for naturalization or the extension of rights and entitlement at all. They may be a part of the community but are excluded from the polity. Under the open-borders scenario, the problems of inequality, injustice, and oppression can only be addressed if all residents – migrant or not – possess the same access to rights and entitlements, independent of their place of birth, ancestry, status, or ability to stay. Residents would then no longer be legally distinguished and treated differently based on their status as migrants ineligible for citizenship, permanent residents with probationary status, or residents with full citizenship.

Currently, states grant and deny migrants rights and citizenship on the basis of carefully calibrated laws and practices of selection and exclusion. In a world of open borders, extending equal rights and citizenship to migrants and non-migrants would be an important step towards realizing the contingent possibility of equality, justice, and freedom from oppression – but it would only be one step. Completing this journey towards equality, justice, and freedom from oppression requires the pursuit of possibilia.

No Border as Possibilia

Let me begin this section by examining some of the contradictions that would emerge if territorial nation states had open borders. One contradiction would be that the open-borders scenario challenges the Westphalian model of territorial sovereignty upon which the current global geopolitical order rests. In Chapter 2, I described an aspect of the border as an instrument of the state to exercise sovereignty. In light of the decreasing ability of individual states to control cross-border trade, fiscal policy, and international affairs, the state's ability to restrict cross-border migration has even been called the "last bastion of sovereignty" (Dauvergne 2007, 2008). Free cross-border mobility would deny sovereign states this legal right – which they claim to possess by virtue of being sovereign – to control migration. The current understanding of state sovereignty seems incompatible with the idea of open borders.

Another contradiction is that the open-borders scenario would fundamentally challenge global capitalism, which is structurally dependent on the divisions that borders create. With the help of borders that lock vulnerable, exploitable, and low-wage labor into the countries of the global south, international companies and corporations are able to produce goods and services at low costs and rake in record profits. Even if this labor manages to cross the border and enter the economies of the global north, border practices ensure that many of these workers remain vulnerable and exploitable. In short, borders enforce an international segmentation of labor that has been integral to global capitalism (Piore 1979; Cohen 1987; Bauder 2006).

Even as the geographical scales of political authority are changing, polities remain territorial in nature. The mere rescaling of territories has not resulted in greater freedom of migration for everyone. For example, the emerging scale of Europe, represented by the European Union, the Eurozone, and the Schengen Area, has open borders internally to European Union citizens who now possess freedom of mobility and the right to reside in any member state, where they can work, vote in municipal elections, and enjoy protection from discrimination based on their nationality. Thus, in light of the recent European financial crisis, Greek, Italian, and Spanish European citizens have responded to high unemployment rates and the dismantling of their welfare systems by moving to booming Germany, where they have practically equal residency and labor market rights as Germans. The freedom of migration within Europe is matched by an identity politics that emphasizes a shared European-ness. In German political and media debates, a migrant from Greece, Italy, and Spain has largely shed the image of the non-belonging foreigner and is now considered a fellow European citizen whose belonging is not questioned in the German media or politics (Bauder 2011). Yet, simultaneously, the exterior border of Europe has hardened. Only privileged non-European nationals are permitted to enter Europe, while less privileged "third-country" nationals are denied entry. The more than 30,000 migrants who died or went missing on their way to Europe since 2000 (Chapter 1), are a gruesome result of the hardening of Europe's external border. If non-European migrants manage to cross the border as temporary migrants, refugees, or irregular migrants, they typically possess no entitlement to citizenship, are treated as non-equals, and often experience discrimination and exploitation. The opening of borders within Europe has merely shifted the denial of freedom of migration and the inequality, injustice, and oppression resulting from bordering practices to a different scale. To prevent this rescaling of deadly and oppressive border practices, the open-borders scenario should apply globally.

However, a world with open borders may not deliver on the promises of equality, justice, and freedom from oppression either. Quite the opposite: under free-market conditions, open borders could fully unleash the brutal forces of capitalism that until so far have been constrained by border controls and migration restrictions (Chapter 3). Under such conditions, open borders would increase global labor competition and drive down wages by pitting migrant and non-migrant workers against each other (Bauder 2006). Open borders could also wreak havoc on national welfare systems by granting migrants access to collective resources to which they have not contributed. National welfare systems in the global north have already endured decades of assault from free-market and small-government reforms and could finally crumble under an open-borders scenario. The economist Milton Friedman astutely explained that you cannot have free immigration and a welfare state – unless the migrants are illegalized and denied legal access to welfare benefits (Friedman 2009). The economic geographer Michael Samers adds that an open-borders world under free-market conditions would be akin to a "neo-liberal utopia" (Samers 2003, 214) in which capital is freely accumulated by a powerful few and the labor of the masses is freely exploited. In the words of Samers' fellow geographer Dan Hiebert (2003, 188–9): "while migration restrictions are based on the protection of privilege, removing those restrictions would not end privilege … such an effort could just as easily lead to mass harm as mass good." In a dialectical contradiction, an open-borders world that promises equality, justice, and freedom also harbors the potential of this world's negation.

Preventing the dystopia described by Samers and Hiebert requires the transformation of the contemporary structures of capitalism. Such a world is located in the realm of possibilia. This realm also entails the transformation of current territorial political organization and belonging, and corresponds to the call for "no border."

As graffiti, the slogan "no border, no nation" adorns trains, buildings, and bridges throughout Europe and North America. Often, this slogan is extended to form rallying cries, such as "no border, no nation, stop the deportation," which links no-border calls to the freedom of migration, or – as in Figure 4.2 – "no nation, no border, fight law and order," which associates no-border calls

Figure 4.2 No Nation, No Border graffiti in Freiburg, Germany, 2015
Source: Photo by Harald Bauder

with demands for a world beyond existing structures of governance. For no-border advocates, the problem is that borders "are the mark of a particular kind of relationship, one based on deep divisions and inequalities between people who are given varying national statuses" (Anderson et al. 2009, 6). In other words, borders are "constitutive" of political categories that grant or deny people rights (Mezzadra and Neilson 2013, xi).

No-border politics therefore focuses on the struggles that occur when people resist the labels of non-citizen, migration, or refugee that borders impose on them. According to Sandro Mezzadra and Brett Neilson (2013, 13–14), these

> border struggles open a new continent of political possibilities, a space within which new kinds of political subjects, which abide neither the logic of citizenship nor established methods of radical political organization and action, can trace their movements and multiply their powers.

In the words of Bridget Anderson and her colleagues (2009, 6), no-border calls challenge "nation states' sovereign right to control people's mobility [and signal] a new sort of liberatory project, one with new ideas of 'society' and one aimed at creating new social actors not identified with national projects." This no-border project pursues ideas of belonging beyond national identity and formal citizenship in a territorial nation state.

Already, citizenship can be understood to be denationalized (Bosniak 2000). Some people may consider themselves "citizens of the world"; others may identify more with cities, towns, regions, or in other territorial or non-territorial terms (Holston 1999; Isin and Nielsen 2008; Isin 2000). Yet, these alternative ideas of belonging may still exclude non-members, and continue to be a source of distinction and inequality. Conventional ways of understanding the world – including the concept of citizenship – will not suffice to achieve the goals of no-border politics. At one point in the future, world civilization may "discover" – in a way that is not foreseeable today – that "the right to free and open movement of people on the surface of the earth is fundamental to the structure of human opportunity" (Nett 1971, 218). This future world civilization may support freedom of migration because social and political practice requires it.

One must be careful not to uncritically embrace no-border calls. Radical transformation always harbors the danger of creating an undesirable outcome. History is full of examples of people's revolutions ending in tyranny, including the Russian revolution against the tsar giving rise to the Stalinist Soviet Union, or the Cuban anti-imperialist revolution to the Castro brothers' rule. If the sovereign territorial state loses the monopoly over controlling cross-border migration, the result may be dystopian. Access to territory could follow a business-transaction model, creating gigantic gated communities for the rich (Torpey 2000, 157). Some states, like Austria, Antigua and Barbuda, Cyprus, Malta, and St. Kitts and Nevis are already offering entry and citizenship to the wealthy in exchange for investments or charitable donations. It would be even less desirable to attain a type of political order exemplified by the recent establishment of an Islamic caliphate on the territory of Syria and Iraq by "a very unprogressive group of Open Borders fanatics" (West 2015) who recognize neither territorial states nor their borders. The reorganization and rethinking of borders and cross-border migration must involve a measured and reflexive dialectical process that includes simultaneous transformations of the political spatial order, practices of migration, and the corresponding ideologies and political principles.

Of course, a no-border world will not be created by academic musing. Rather it will emerge from the collective practices involving the people who have been denied migration or who are illegalized and disenfranchised. These practices begin with a negation: the rejection of borders, territorial statehood, and nationhood, as well as derived labels, such as foreigner or migrant, because these labels oppress the people they identify (Sharma 2003; Wright 2003). These collective practices, however, can also achieve more. They have the potential to create new political categories and "novel forms of political subjectification" (Nyers 2010, 137).

Solidarity is central to no-border politics and the creation of new political actors. Solidarity can involve loyalty, empathy, moral principle, and self-serving utility (Kapeller and Wolkenstein 2013). In the context of social activism, it often expresses a "community of interests, shared beliefs and goals around which to unite" (Hooks 2000, 67). It also entails listening to other people and understanding their experiences, needs, and desires; solidarity affirms community by "creating bonds of love, trust, respect, compassion, and mutual aid" (Walia 2013, 269). My understanding of solidarity follows a Hegelian tradition, which suggests that a new political consciousness emerges from social practice and action. By *acting* in solidarity, citizens, migrants, and other people participate in the dialectical formation of political consciousness. In this way, no-border politics follows a long tradition of critical practice, which ranges from organizing the working class in the 19th century (Marx and Engels 1953; 1967 [1848]) to the "consciousness-raising" (Pratt and Rosner 2006, 15) among today's marginalized migrant communities. This tradition of critical practice pursues social and political transformation through acts of solidarity.

Although no-border politics pursues the elusive goal of possibilia, it is not detached from the contemporary world and everyday affairs. Quite to the contrary. By rejecting existing structures of oppression and the violence these structures inflict, the practical politics of no border remain deeply rooted in the problems created by today's nation states, territorial membership, and global capitalism (Burridge 2014). Thus, no-border politics is "not utopian. It is in fact eminently practical and is being carried out daily" (Anderson et al. 2009, 12). I concur that no-border politics is "not utopian" – not only because this politics is already being carried out, but also because no-border politics refrains from articulating concrete alternatives (i.e. contingent possibilities) typically associated with the concept of utopia. It is neither possible nor would it be desirable to define a no-border world as a concrete utopia.

Conclusion

Calls for open borders and no border reject contemporary border and migration controls. Open-borders and no-border calls, however, refrain from describing the circumstances under which freedom of migration is supposed to occur. In this way, they avoid the pitfall of articulating concrete alternative futures using the very concepts and ways of understanding the contemporary world that critical theorists, such as Theodor Adorno, and activists, such as Harsha Walia, are seeking to overcome.

Yet, I will continue to explore the contingent possibility of open borders in Part II of the book. Under current circumstances, the territorial nation state provides the practical resources that make freedom of migration "feasible." Christian Matheis and I have recently described this feasible possibility as a "mesolevel" intervention, between business-as-usual, pork-barrel politics and revolutionary transformation (Matheis and Bauder 2016). If we assume that the bordered territorial state will not go away anytime soon, what should be our way forward to deliver on the promises that freedom of migration makes? I pursue this question in Chapter 5. We must be mindful, however, that any positively articulated future using existing concepts and ways of thinking about the world has an inherent limitation: it inevitably denies possibilia that emerges under not-yet existing circumstances and ways of thinking. The contingent possibility of open borders and the possibilia of no border stand in dialectical opposition to each other.

The politics of no border and the vision of open borders may be contradictory but they are not antagonistic. In fact, the way forward must involve both; neither the open-borders possibility nor the no-border possibilia are sufficient by themselves. Rather, the tension between the open-borders vision and no-border politics marks a critical moment in the dialectic towards freedom of migration. Both notions of open borders and no border play important parts within this dialectical progression.

Bloch and other philosophers have used the metaphor of the "horizon" to illustrate how the dialectic towards the future unfolds: the unknowable possibilia that emerges on the metaphorical "horizon" inspires today's social and political practices that shape the not-yet-existing future (Bloch, 1985 [1959], 328–34). Located halfway between us and this fuzzy horizon is the contingently possible. Like the rainbow – this symbol of hope – the contingent possibility of open borders provides a concrete reference point on the path towards the possibilia of freedom of migration. However, also like the rainbow that keeps moving away as we approach it, the concrete vision of an open-borders world can only serve as a provisional inspiration until new material circumstances render it redundant. Conversely, the no-border narrative lies in the realm of possibilia and cannot be seen from our current location. Part II of the book grapples with these issues.

References

Adorno, Theodor W. 1966. *Negative Dialektik*. Frankfurt am Main: Suhrkamp Verlag.

Adorno, Theodor W. and Ernst Bloch. 2014. "Möglichkeiten der Utopie heute (1964)." Youtube, April 29. Accessed January 29, 2016. https://www.youtube.com/watch?v=oRz3BnpqmhE.

Alldred, Pam. 2003. "No Borders, No Nations, No Deportations." *Feminist Review* 73: 152–157.

Anderson, Bridget, Nandita Sharma, and Cynthia Wright. 2009. "Why No Borders?" *Refuge* 26(2): 5–18.

Bauder, Harald. 2006. *Labor Movement: How Migration Regulates Labor Markets*. New York: Oxford University Press.

Bauder, Harald. 2011. *Immigration Dialectic: Imagining Community, Economy and Nation*. Toronto: University of Toronto Press.

Bauer, Otto. 1907. *Die Nationalitätenfrage und die Sozialdemokratie*. Vienna: Verlag der Wiener Volksbuchhandlung Ignaz Brand.

Best, Ulrich. 2003. "The EU and the Utopia and Anti-Utopia of Migration: A Response to Harald Bauder." *ACME* 2(2): 194–200.

Bloch, Ernst. 1985 [1959]. *Das Prinzip Hoffnung*. Frankfurt/Main: Suhrkamp.

Bosniak, Linda S. 2000. "Citizenship Denationalized." *Indiana Journal of Global Legal Studies* 7(2): 477–509.

Bosniak, Linda S. 2007. "Being Here: Ethical Territorial Rights of Immigrants." *Theoretical Inquiries in Law* 8(2): 389–410.

Burridge, Andrew. 2014. "'No Borders' as a Critical Politics of Mobility and Migration." *ACME* 13(3): 463–470.

Carens, Joseph H. 2000. "Open Borders and Liberal Limits: A Response to Isbister." *International Migration Review* 34(2): 636–643.

Casey, John P. 2009. "Open Borders: Absurd Chimera or Inevitable Future Policy?" *International Migration* 48(5): 14–62.

Cohen, Robin. 1987. *The New Helots: Migrants in the International Division of Labour*. Aldershot: Avebury.

Dauvergne, Catherine. 2007. "Citizenship with a Vengeance." *Theoretical Inquiries in Law* 8(2): 489–506.

Dauvergne, Catherine. 2008. *Making People Illegal: What Globalization Means for Migration and Law*. New York: Cambridge University Press.

Department of Homeland Security. 2015. "Policy Manual, Volume 12, Citizenship and Naturalization." US Citizenship and Immigration Services. Last modified November 10. Accessed December 12, 2015. http://www.uscis.gov/policymanual/HTML/PolicyManual-Volume12.html.

Engels, Friedrich. 1971 [1880/1882]. *Die Entwicklung des Sozialismus von der Utopie zur Wissenschaft*. Berlin: Dietz Verlag.

Friedman, Milton. 2009. "Illegal Immigration." Youtube, December 11. Accessed October 18, 2015. https://www.youtube.com/watch?v=3eyJIbSgdSE.

Harvey, David. 2000. *Spaces of Hope*. Berkley, CA: University of California Press.

Harvey, David. 2005. *A Brief History of Neoliberalism*. Oxford: Oxford University Press.

Hawel, Marcus and Gregor Kritidis. 2006. "Vorwort." In *Aufschrei der Utopie: Möglichkeiten einer anderen Welt*, edited by Marcus Hawel and Gregor Kritidis, 7–8. Hannover: Offizin-Verlag.

Hiebert, Daniel. 2003. "A Borderless World: Dream or Nightmare?" *ACME* 2(2): 188–193.

Holston, James, ed. 1999. *Cities and Citizenship*. Durham, NC: Duke University Press.

Hooks, Bell. 2000. *Feminist Theory: From Margin to Centre*. London: Pluto Press.

Isin, Engin F., ed. 2000. *Democracy, Citizenship and the Global City*. London: Routledge.

Isin, Engin F. and Greg M. Nielsen, eds. 2008. *Acts of Citizenship*. New York: Zed Books.

Kapeller, Jakob and Fabio Wolkenstein. 2013. "The Grounds of Solidarity: From Liberty to Loyalty." *European Journal of Social Theory* 16: 476–491.

Levitas, Ruth. 1990. *The Concept of Utopia*. Syracuse, NY: Syracuse University Press.

Mannheim, Karl. 1952 [1929]. *Ideologie und Utopie*. Frankfurt am Main: Schulte-Bulmke.

Marcus, Ruth Barcan. 1975–6. "Dispensing with Possibilia." *Proceedings and Addresses of the American Philosophical Association* 49: 39–51.

Marx, Karl and Friedrich Engels. 1953. *Die Deutsche Ideologie*. Berlin: Dietz Verlag.
Marx, Karl and Friedrich Engels. 1967 [1848]. *The Communist Manifesto*. Harmondsworth: Penguin.
Marx, Karl. 1982 [1848]. "Der 'Débat social' vom 6. Februar über die Association Démocratique." In *Werke Volume 4*, by Karl Marx and Friedrich Engels, 511–513. Berlin: Dietz Verlag.
Matheis, Christian and Harald Bauder. 2016. "Possibility, Feasibility and Mesolevel Interventions in Migration Policy and Practice." In *Migration Policy and Practice: Interventions and Solutions*, edited by Harald Bauder and Christian Matheis, 1–16. New York: Palgrave Macmillan.
Mezzadra, Sandro and Brett Neilson. 2013. *Border as Method: or, the Multiplication of Labor*. Durham, NC: Duke University Press.
More, SirThomas. 1997 [1516]. *Utopia*. New York: Dover Publications.
Nett, Roger. 1971. "The Civil Right We Are Not Ready For: The Right of Free Movement of People on the Face of the Earth." *Ethics* 81: 212–227.
Nute, Donald. 1998. "Possible Worlds without Possibilia." In *Thought, Language, and Ontology*, edited by Francesco Orilia and William J. Rapaport, 153–167. Dordrecht: Kulwer Academic Publishers.
Nyers, Peter. 2010. "No One Is Illegal between City and Nation." *Studies in Social Justice* 4(2): 127–143.
Paasi, Anssi. 2009. "Bounded Spaces in a 'Borderless World': Border Studies, Power and the Anatomy of Territory." *Journal of Power* 2(2): 213–234.
Piore, Michael. 1979. *Birds of Passage: Migrant Labor and Industrial Societies*. Cambridge: Cambridge University Press.
Pratt, Geraldine and Victoria Rosner. 2006. "The Global and the Intimate." *Women's Studies Quarterly* 34(1/2): 13–24.
Quarta, Cosimo. 1996. "Homo Utopicus: On the Need for Utopia." *Utopian Studies* 7(2): 153–166.
Rorty, Richard. 1991. *Objectivity, Relativism, and Truth: Philosophical Papers*, Volume 1. Cambridge: Cambridge University Press.
Samers, Michael. 2003. "Immigration and the Spectre of Hobbes: Some Comments for the Quixotic Dr. Bauder." *ACME* 2(2): 210–217.
Sassen, Saskia. 2006. *Territory, Authority, Rights: From Medieval to Global Assemblages*. Princeton, NJ: Princeton University Press.
Sharma, Nandita. 2003. "No Borders Movements and the Rejection of Left Nationalism." *Canadian Dimensions* 37(3): 37–39.
Soysal, Yasmin N. 1994. *Limits of Citizenship: Migrants and Postnational Membership in Europe*. Chicago, IL: University of Chicago Press.
Torpey, John. 2000. *The Invention of the Passport: Surveillance, Citizenship and the State*. Cambridge: Cambridge University Press.
Voltolini, Alberto. 1994. "Ficta versus Possibilia." *Grazer Philosophische Studien* 48: 75–104.
Walia, Harsha. 2013. *Undoing Border Imperialism*. Oakland, CA: A. K. Press.
Walzer, Michael. 1983. *Spheres of Justice: A Defense of Pluralism and Equality*. Oxford: Martin Robertson.
Wells, H. G. 1959 [1905]. *A Modern Utopia and Other Discussions: The Works of H. G. Wells*, Atlantic Edition, Volume IX. London: T. Fischer Unwin.
West, Ed. 2015. "Isis Are Just Very Un-progressive Open Border Fanatics: We Need an Atatürk to Fight Them." *Spectator*, February 20. Accessed February 20, 2016.

http://blogs.spectator.co.uk/2015/02/isis-are-just-very-un-progressive-open-border-fanatics-we-need-an-ataturk-to-fight-them/.

Wilde, Oscar. 1891. *The Soul of Man under Socialism*. Accessed February 17, 2016. https://www.marxists.org/reference/archive/wilde-oscar/soul-man/.

Wright, Cynthia. 2003. "Moments of Emergence: Organizing by and with Undocumented and Non-Citizen People in Canada after September 11." *Refuge* 21(3): 5–15. Accessed January 7, 2013. http://pi.library.yorku.ca/ojs/index.php/refuge/article/viewFile/23480/21676.

Part II

Solutions

> We have to stop talking about integration … democracy is not a country club. Democracy means that everyone has the right to determine for themselves and with others how they want to live together … If integration means anything, it is that we are all in this together!
>
> Kritnet (2011)

> Let us fight to free the world, to do away with national barriers, to do away with greed, with hate and intolerance.
>
> Charlie Chaplin (1940)

In Part I of this book, I offered a diagnosis of the problematic nature of international borders and cross-border migration. In Part II, I explore practical and far-sighted solutions of how freedom of migration can indeed lead to greater equality, justice, and freedom from oppression. Part II follows up on the ideas of contingent possibilities and possibilia that I introduced in the preceding chapter.

In addition, the narrative progresses thematically. While Part I dealt with borders and cross-border migration, Part II focuses on themes of citizenship and territorial belonging. These two themes are closely related to each other: if borders are open, the inevitable question is if and how migrants would be permitted to belong to their chosen communities, and become members with the entitlement – as Kritnet demands – to decide how they want to live together with others.

Chapter 5 begins with the observation that citizenship practices continue to exclude migrants who were able to cross borders. Therefore, I investigate if and how citizenship based on residence – rather than birth privilege – can offer the possibility of open borders without alienating migrants in their countries of destination. Although some readers may consider such a possibility to be far removed from current citizenship practices in most countries, it does assume that the basic structure of the nation state with its territorial borders does not change. In this way, citizenship based on residency represents a contingent possibility.

In Chapter 6, I take this narrative a step further. In particular, I explore a different geographical scale of belonging, one that is not defined by the nation state but by the city. This chapter applies the work of forward-looking urban theorists to the empirical context of existing urban communities in North America that resist the disenfranchisement of migrants, offering "sanctuary" for illegalized migrants.

Finally, in Chapter 7, I shift gears again and explore the possibilia of a radically transformed society that realizes freedom of mobility. Rather than providing a blueprint for such a society, I contemplate the circumstances that may bring about this transformation. I use the practices and actions of the no-border movement to illustrate pathways towards such a real possibility.

References

Chaplin, Charlie. 1940. The Great Dictator. Charles Chaplin Film Corporation.

Kritnet (Netzwerk kritische Migrations und Grenzregimeforschung). 2011. "Democracy Not Integration." Accessed November 28, 2015. http://demokratie-statt-integration.kritnet.org/demokratie-statt-integration_en.pdf.

5 Mobility and Domicile

> Birthright-based membership rules and the legal kinship paradigms sustaining these, are the root cause of systemic violence and inequality within and among political societies.
>
> Jacqueline Stevens (2010, 75)

People are not treated equally at international borders. When international travelers step off the plane at an airport in the USA and approach the customs and immigration area, they are immediately sorted based on their citizenship. As non-citizens may be required to line up behind a long cue, they can watch the US citizens who sat next to them on the same plane zip through immigration more quickly. A similar scenario awaits travelers arriving at international airports in Europe's Schengen Area. For many travelers, this unequal treatment amounts to small inconveniences; they may miss their connecting flight or arrive in their hotels later than they had hoped. Most people who do not have the state's permission to enter the country will have been denied boarding the plane at the departure airport. Occasionally, however, an immigration officer at an international airport refuses a traveler entry. Although this traveler may have lived in the USA or Europe for years, and have a family, home, and job there, an expired visa or work permit or a denied refugee claim may render this person "inadmissible."

Citizenship is vital for migrants: it stipulates whether a person has the right to cross an international border; whether someone requires a visa or permit, or whether crossing the border legally is impossible. After migrants have crossed the border, access to status or that country's citizenship regulates whether they are permitted to remain in the country and what rights and legal entitlements they possess while they live there.

Over the last two decades, the concept of citizenship has received widespread attention among scholars. Researchers have explored various aspects of citizenship, ranging from policies and practices of inclusion to the way in which people enact their political claims. Yet, what matters for migrants first and foremost is the passport. This *formal* aspect of citizenship is the topic of this chapter.

Early political theorists like Jean-Jacques Rousseau (2003 [1762], 31) presupposed that a political community is organized on the basis of territory, which sustains the citizens "who make the state." In this tradition, political theorist Hannah Arendt (1968, 81) declared almost half a century ago: "Nobody can be a citizen of the world as he [sic] is the citizen of his [sic] country." For Arendt, citizenship defines legal membership in a territorial nation state. Law scholar Linda Bosniak (2000, 456) agrees: "Citizenship is almost always conferred by the nation state, and as a matter of international law, it is nation-state citizenship that is recognized and honored." Through formal citizenship, nation states grant legal equality and political, social, and economic rights, provide equal labor market access, redistribute social and economic resources, promise protection against oppression, and permit both entry onto state territory and the right to remain there.

Formal citizenship is a mechanism of inclusion in and exclusion from a national community. Migrants tend to be de-facto members of the community in which they reside; they typically pay taxes and participate in social and civic life. Yet, they often lack access to formal citizenship. This exclusion from citizenship is an underlying reason for temporary foreign workers to be treated unequally compared to workers who possess citizenship, even though they may live and work in the same country. For example, Canada's Live-In Caregiver Program has created a workforce of mostly Filipina women, who were not given the option of bringing their own children and families if they wanted to come to Canada to take care of the children and families of Canadians. While they work and live in Canada, they lack the security and protection that Canadian citizens take for granted. When all Canadian temporary foreign workers programs are combined, the size of the temporary foreign workforce in Canada now exceeds the number of permanent immigrants who annually arrive in Canada. In this case, the Canadian state uses citizenship as a strategic tool to create a population that is formally excluded from membership in Canadian society and that will therefore need to accept working conditions which citizens would deem unacceptable.

The Gulf states are an example par excellence of how citizenship can be used to strategically exclude migrants who live in the country. Of the total population of the United Arab Emirates (UAE), only slightly more than one out of ten is considered a local resident – that is, a full citizen, passport holder, or a Bedouin. Citizenship and a passport come with considerable employment privileges and access to social services. The remaining population – almost nine out of ten – consists of expatriates, who are organized into a hierarchy that privileges Westerners over South and East Asian laborers. This citizenship practice – which has its origin in British geostrategic interests that predates UAE federation – has been an effective tool to import knowledgeable experts as well as vulnerable manual labor on which the country depends (Jamal 2015). Human Rights Watch (2014, 224–5) reports the following situations, which illustrate the circumstances facing migrant workers in UAE:

> In May, hundreds of workers at a site in Dubai went on strike demanding better pay and conditions. After the two-day strike, immigration officials issued at least 40 deportation orders.
>
> UAE labor law excludes domestic workers, almost exclusively migrant women, denying them basic protections such as limits to hours of work and a weekly day off.
>
> Without citizenship, these workers are unable to claim basic labor rights and protections.

The lack of access to citizenship has also been a major cause of the illegalization of migrants. The International Organization for Migration (IOM 2014) estimates that at least 50 million migrants in the world do not possess legal status in the country in which they are located. Illegalization occurs in several ways. One way is that a person crosses the border without legal permission to do so. This situation applies to a large portion of the illegalized population in the USA, where an estimated 11.3 million illegalized migrants lived in 2014, according to a report published by the Pew Research Center. The vast majority of these illegalized immigrants are long-term residents of the USA and have lived there for more than a decade; almost four out of ten (38 percent) live with children who were born in the USA (Passel and Cohn 2015). Another form of illegalization occurs when people enter a country as refugees or asylum seekers and go underground after their claims were rejected. Finally, a person may enter a country as a visitor, student, or worker but stay in the country past the expiry date of the visa or work permit. In either of these cases, the lack of citizenship of the country in which someone resides tends to render a person "illegal" and subsequently vulnerable to abuse and exploitation (Bauder 2013).

The underlying reason of why people who reside in a national territory can be treated differently lies in the way states grant citizenship. They typically do this at birth, based on the national territory in which a person is born or on the citizenship of the parents. Then, the state assumes that people do not migrate between countries or national communities and therefore will keep this citizenship for the remainder of their lives. Other ways of granting citizenship, such as through naturalization, is assumed to be an exception. In this way, current citizenship practice reproduces the birth privilege of non-migrants. As a solution to this problem, I suggest implementing the *domicile* principle of citizenship.

Let me first clarify the terminology. "Citizenship principle" refers to the mechanism based on which individuals acquire formal citizenship and become formal members of the polity. The term "domicile" has its roots in the Latin noun *domicilium*, which can be translated as household, habitation, home, or residence. Correspondingly, the principle of domicile refers to citizenship based on "effective residence" (Hammar 1990, 76). When legal and citizenship scholars discuss this principle of citizenship, they sometimes use the term *jus domicilii*, which is Latin for "law of residence." This citizenship law thus postulates that people residing in a territorial state have the right to citizenship, independent of a state's or a ruling elite's efforts to exclude some people from formal membership based on their migration history. In other words, people are citizens of the country in which they reside, no matter where or to whom they were born.

Below, I develop a practical argument in support of domicile-based citizenship. This argument presupposes the continuation of prevalent political structures that exist today. In particular, I assume that three conditions will not change: first, nation states will continue to be territorially organized and geographically bordered; second, legal membership in this territorial state will be regulated through formal citizenship; and third, citizenship will be framed in universal terms so that the same rules of membership apply to everyone. The pragmatic observer may see the persistence of these three conditions as common sense. Although more critical and radical scholars have questioned the association between citizenship and the territorial nation state and critiqued the decontextualized application of universal citizenship (e.g. Bosniak 2000; Cresswell 2006; Isin 2012), in this chapter I choose to overlook these concerns. Rather, the practical argument I am making in this chapter resembles what I called a "contingent possibility" in Chapter 4. It does not consider the possibilia of a radically transformed world, in which the territoriality of the state, formal citizenship, and international borders may have been abolished. Instead, in this chapter I recognize that contemporary territorial states and citizenship possess considerable capacities to provide rights to people and to guarantee equality among human beings within the territorial polity. I propose, however, to change the way citizenship is granted, advocating the domicile principle of citizenship as a practical tool to include migrants in the polity. This tool is of increasing relevance as more and more people around the globe are mobile, societies are more transnational than they have ever been, and large numbers of migrants are excluded from membership in the communities in which they live.

Principles of Citizenship

The citizenship principles we hear most about are *jus sanguinis* and *jus soli*. Both are birthright citizenships. *Jus sanguinis* refers to the acquisition of citizenship through blood lineage (*sanguis* = blood). According to this principle, a child possesses the citizenship(s) of its parents. This principle has its roots in ancient Greece. It was established in the middle of the 5th century BC in Athens, after aristocratic clans abused existing citizenship practices and expanded their political influence by granting citizenship to foreigners (Bauböck 1994, 38). By adopting inherited citizenship, the membership of the political community became immune "against arbitrary decisions by authorities" (Bauböck 1994, 45). Later, in the context of migration, *jus sanguinis* also had advantages for colonizing and emigration countries, because it enabled states to bond colonial and expatriate communities abroad (Castles and Davidson 2000, 85). It entitled, for example, German colonizer communities in Eastern Europe to maintain their ancestral ties and their right to return to German territory, although they had been absent from that territory for generations. After the Second World War, these German nationals who lived in Eastern Europe under oppressive communist regimes retained their national membership and were accepted as nationals in West Germany.

Jus soli refers to the acquisition of citizenship based on place of birth (*solum* = soil, ground, country). According to this principle, a child is a citizen of the country in which it was born. *Jus soli* too has a long history. For example, *jus soli* was applied under European feudalism when feudal lords reigned over land and "anybody born within it" (Bauböck 1994, 35). *Jus soli* has been widely adopted by settler countries, such as Canada and the United States, to tie the descendants of newcomers to the state.

In the United States *jus soli* "birthright" citizenship has been under attack recently. The critics of automatic *jus soli* citizenship would prefer to deny citizenship to US-born children of illegalized parents living on US soil, while they want to continue granting *jus soli* citizenship to children of parents who also possess citizenship. These critics apparently want to shift US citizenship practice from *jus soli* to *jus sanguinis*. They are obviously not interested in abolishing birthright, but in restricting it for the privileged.

Despite their fundamental differences, *jus domicilii, jus sanguinis*, and *jus soli* also share important characteristics. Both *jus domicilii* and *jus soli* are territorial, which means that citizenship is tied to the geographically bounded territory of the state. Conversely, *jus domicilii* and *jus sanguinis* are both indifferent to the location of a person's birth. The difference that matters most to my argument is that both *jus sanguinis* and *jus soli* grant citizenship as a birthright, while *jus domicilii* does not. Under *jus soli*, persons who were not born in the country can be excluded from membership in the national community. This situation applies to temporary foreign workers and illegalized migrants. Conversely, under the *jus sanguinis* principle, generations of foreigners may live in a country without entitlement to citizenship. This situation occurred in Germany, where the German-born children of foreign "guest workers" – and the children of these children – were for a long time not granted German citizenship, despite generations of residence in Germany.

The combination of *jus soli* and *sanguinis* principles does not always resolve the contradiction that the residents in a country are excluded from citizenship. For example, a German by descent (with *jus sanguinis* German citizenship) born in Canada (with *jus soli* Canadian citizenship) still does not possess an entitlement to citizenship in a third country in which this person may reside. Conversely, *Jus domicilii* would not exclude anyone based on whether she or he was born to the "wrong" parents or in the "wrong" country. In this way, domicile-based citizenship accommodates people who migrate between communities and territories. It treats people as members based on the territory in which they reside, independent of the circumstances of their birth.

The domicile principle of citizenship has been called the "missing link" (Gosewinkel 2001, 29) between *jus sanguinis* and *jus soli*. In fact, in practice most countries combine all three citizenship principles in some way. Countries with a strong *jus soli* tradition typically also grant citizenship to the offspring of its citizens (although this rule may no longer be applied after a few generations have been absent from state territory), and they naturalize immigrants based on residency criteria. Similarly, European countries like Germany with a *jus sanguinis* tradition increasingly grant citizenship to children born on their national territory, provided that the parents fulfill certain residency requirements. These European countries have recently incorporated *jus domicilii* and *jus soli* elements into their citizenship legislation. Empirical research furthermore shows that public opinion across a diverse range of countries favors combinations of *jus sanguinis, jus soli*, and *jus domicilii* (Levanon and Lewin-Epstein 2010; Raijman et al. 2008; Ceobanu and Escandell 2011). My argument expands beyond the state practices and public opinion polls that favor combining the different citizenship principles. Rather, I advocate elevating the domicile principle over other principles that frame citizenship as a birthright.

Domicile as Historical Practice

The domicile principle of citizenship is not a new idea or practice. Similar to *jus soli* and *jus sanguinis*, it has a long history. In this section, I explore how the domicile principle was applied in the past to include people who migrated between state territories.

When legal scholar Rolf Grawert (1973) examined the origin of nationality and citizenship in Europe, he found that the feudal order applied not only *jus soli* but also the domicile principle as a way of bonding people to territory in which they were not born. In the 16th and 17th centuries, the legal answer to the question of how a person becomes a subject of a feudal lord was: "domicilum facit subditum" (domicile makes the subject) (Grawert 1973, 79). Correspondingly, legal documents in medieval Europe used various Latin terms related to *domicilium* to articulate the territorial belonging of legal subjects. This medieval application of the domicile principle combined Roman law and Catholic Church law, according to which domicile in a territory refers to both being a citizen of a jurisdiction and a resident in a particular place.

As feudalism came to an end in Europe, the domicile principle persevered. In the wake of the French Revolution, it was affirmed as an important citizenship principle. Social and political theorist Rainer Bauböck in fact suggests that the following passage, taken from the 1793 French Constitution, could be "the most radical formulation of *jus domicili* in history" (Bauböck 1994, 32):

> every foreigner who has completed his 21st year of age and has been resident in France for one year and lives from his labour or acquires a

> property or marries a French spouse or adopts a child or nourishes an aged person ... is admitted to the exercise of French citizenship.
>
> (translated by Bauböck 1994, 50)

French citizenship law clearly followed the domicile principle – but under conditions: there was a one-year waiting period and a candidate needed to engage in productive or reproductive labor.

On the other side of the Rhine river, in Germany, the domicile principle was also an important legal practice. In the 19th century, German territory was fragmented into dozens of independent states and cities. In order to prevent statelessness among people who migrated within German territory, the independent states and cities committed to treating migrants as their own and naturalized them after certain periods of residency had elapsed. According to Grawert (1973, 75) this legal practice effectively implemented the domicile principle (*Domizilprinzip*) of citizenship. Political scientist Simon Green (2000, 108) agrees that "most German states preferred the principle of residence during the first half of the nineteenth century." At the same time, however, *jus sanguinis* citizenship was also practiced by the German states and cities. Eventually, the Nationality Law of 1913 tied German citizenship throughout the German empire more firmly to descent, replacing domicile as a dominant citizenship principle (Gosewinkel 2001).

In more recent times, the domicile principle continues to guide legal practice. The place of effective residence is an important criterion that courts use to decide on the dominant nationality of persons with multiple citizenships (Hammar 1990). In addition, domicile is usually the central criterion for purposes of taxation, especially in cases whereby a person has incomes in multiple countries or is mobile between countries. For the purpose of regulating matters of international taxation, the Organisation for Economic Co-Operation and Development (OECD) therefore establishes criteria for measuring domicile that apply to persons in such situations. According to the OECD's *Model Tax Convention*, domicile is based on the location of an individual's "permanent home," "centre of vital interests," "personal and economic relations," a "habitual abode," and formal nationality (OECD 2015, article 4). In the European Union, persons qualify for social security either in the country in which they are employed or where they are "habitual resident," measured by criteria such as family ties, duration and continuity of presence, the practice of non-remunerated activities, source of income, permanence of housing situation, and where they pay taxes (European Commission 2014).

In the case of taxation and social security, the principle of domicile is used to establish the jurisdiction in which a person has legal responsibilities and entitlements. It does not apply to citizenship per se. Writing from a historical perspective, Grawert (1973, 224, my translation) observes that, unlike today, in Central Europe "the voluntary long-term establishment of residence in a state territory was definitely a legitimate reason to grant citizenship. Up until the 18th Century, neither states nor cities had difficulties accepting this practice."

By the 1970s, West Germany no longer followed this practice, because policy and law makers wanted "to control the undesired naturalization of foreigners" (224). In other words, the domicile principle was suspended for the very purpose of excluding migrant populations, such as the so-called guest workers who resided in West Germany. Conversely, as historical law of citizenship or as a contemporary criteria for taxation and social security, the domicile principle serves territorial political entities to *include* mobile populations. This function is central when I discuss this principle as an alternative to existing citizenship practices and policies below.

Domicile as a Right

The arguments in favor of applying the domicile principle to the realm of citizenship share similarities with the arguments for open borders that I presented in Chapter 3. In particular, these arguments can be made from a range of philosophical positions. From a liberal perspective, the domicile citizenship principle is highly appealing. In the words of legal scholar Yishai Blank (2007, 425), the principle of residence – which is synonymous to the principle of domicile – is "supremely liberal: voluntary, rational, and justifiable; one elects in which locality to live, contributes to it through her taxes and/or activities, is granted membership in this community, and is given equal rights that follow this membership status." Most importantly, domicile-based citizenship rejects birth privilege. Rather than being born to parents who are citizens and in a territory of which one is a citizen, the domicile principle is based on a person's choice – in this case, the choice of residing in a particular country. From this liberal perspective the domicile principle is also equitable because it extends membership to all residents, independent of whether they were born in a country or whether they chose to migrate there. In this way, the domicile principle of citizenship is democratic, as international relations and legal scholar Dora Kostakopoulou observes: "democratic decision making and the flourishing of a political community require the involvement of all the community – not simply of a segment of it" (Kostakopoulou 2008, 126). The domicile principle extends democratic political participation to all de-facto members of the territorial community.

From a political-economy perspective, the domicile principle of citizenship also has merit. This principle prevents legal distinctions among workers who are residing in the same territory. Currently, members of the community who are excluded from citizenship often make greater economic contributions and personal sacrifices than citizens, while they receive fewer social benefits and less legal protection. In short, they do not receive their fair share of the pie, given the contributions they are making. The *domicile* principle of citizenship would tackle this injustice. All de-facto members within a society would obtain citizenship and receive equal treatment in the labor market. Legal status would no longer render migrants more vulnerable and exploitable than others.

Carly Austin and I (2012) examined what would happen if the domicile principle were applied to the case of temporary foreign workers who are denied a pathway to citizenship. We found that enacting this citizenship principle would go a long way towards addressing the exploitation of workers who participate in Canada's temporary foreign worker programs. Domicile-based citizenship would enable temporary foreign workers to compete in the labor market on a level playing field with other workers. In addition, domicile-based citizenship would give migrant workers, who desire to stay in the country, the right to do so. They would no longer be forced to go underground after their temporary status expires if they decide to continue living in the communities to which their labor contributes.

The domicile principle of citizenship has a lot going for it from other perspectives as well. It would help achieve a free market economy, in which labor markets are no longer distorted by the differential treatment of workers. Currently, some workers possess all the rights and protections that come with citizenship while others are denied many of these rights by virtue of not possessing citizenship. Domicile-based citizenship would also make it harder for states to discriminate based on gender and race. Under conditions of domicile-based citizenship, residents would no longer be separated into categories of citizens, privileged migrants who are granted status and the prospect of citizenship, and disadvantaged migrants who are often women and racialized people.

All the arguments I presented above imply that domicile-based citizenship should be a fundamental right rather than a matter of state discretion. Already, naturalization practices in many countries of the global north emulate the domicile principle (Hammar 1990, 76). These practices provide long-term residents the opportunity to acquire formal citizenship. Especially traditional immigration countries, such as Australia, Canada, and the USA, encourage immigrants to apply for citizenship when they meet certain conditions, typically related to the length of permanent residency, basic knowledge of the country's history and system of governance, clearing a criminal background check, and demonstrating (or expressing) their loyalty to the state. According to the US Department of Homeland Security (2015), the United States naturalized almost 780,000 foreign-national residents in 2013 alone. Countries that historically favored the *jus sanguinis* principle of citizenship, too, naturalize many long-term and second-generation foreign residents. However, naturalization is usually not a fundamental right, as Stephen Castles and Alastair Davidson point out:

> Naturalization is a *discretionary act* by the state, one usually carried out by the executive in the form of the head of state, a government minister or a bureaucracy … Apart from certain exceptional circumstances, immigrants have *no entitlement* to naturalization and are mere objects of decisions from above.
>
> (Castles and Davidson 2000, 86, emphasis in original)

Thus, residing legally in the country is not synonymous with being formally considered a member of the national community. Furthermore, entry and thus residence in a state's territory is conditional on selective immigration policies and restrictive residency criteria that clearly do not treat all people equally. People who do not meet the immigration or residency requirement, including temporary residents, are denied the possibility of domicile-based naturalization. The current policies of both traditional and non-traditional immigration countries are designed to carefully select migrants who are permitted to enter the country in the first place, or assess refugee claims to determine if a person can stay. Thereafter migrants and refugees are placed under a period of probation during which they are vulnerable and possess an intermediate-level status. Only then are they eligible for naturalization.

The differential treatment of people based on their citizenship, residency status, and migration history also affects other entitlements. The sociologist Thomas Faist (1995) examined the degree to which migrants in Germany and the United States possess access to economic welfare, security, and social well-being. He found that access varies between people who are residing in the country depending on their status as illegalized migrants, refugees, temporary residents, or permanent residents. In other words, domicile-based social rights – which citizens are entitled to – are not extended to the same degree to non-citizen residents.

The problem with contemporary state practices is precisely that legal status rather than territorial presence defines migrants' access to citizenship and entitlements. In keeping with the spirit of the domicile principle, citizenship should be a right for *all* de-facto residents in a political territory. However, contrary to the spirit of the domicile principle, many states impose visa and immigration restrictions that limit the period foreigners are permitted to stay on state territory, to prevent the possibility that migrants acquire residency-based rights and become eligible for citizenship. For example, the strict, four-year time limit imposed on temporary foreign workers in Canada is designed precisely to force these workers out of the country before they can claim a moral right to domicile citizenship. The Canadian state appears to have learned a disturbing lesson from West Germany's "guest" workers program of the 1950s to 1970s. The foreign workers from Southern Europe and the Mediterranean region were initially intended to be temporary "guests" in West Germany, but they were permitted to renew their employment contracts, thus acquiring the right to stay and eventually reunite with their families in Germany. The domicile principle of citizenship entails that de-facto residents cannot be forced to leave, but have the right to remain in the country and acquire full formal citizenship. This principle must apply to everyone and "cannot be conferred selectively on some residents and denied to others" (Austin and Bauder 2012, 31).

Domicile as Contingent Possibility

Several practical questions emerge in respect to the implementation of the domicile principle of citizenship. One question is this: what criteria exactly should define "domicile"? In other words, how should we determine that a migrant is a resident? Kostakopoulou (2008, 114) argues that citizenship should be extended to immigrants if they intend "to reside in a country indefinitely." This does not mean that a person would be bonded for life to a particular territory. Rather, what counts is the *intention* to stay permanently; and if this intention changes, then citizenship would expire. This argument, however, neglects that temporary residents, too, have a moral claim to citizenship because they are making significant economic and social contributions to the communities in which they reside. In fact, Kostakopoulou's argument seems inconsistent. In the same way that a resident with the intention to stay permanently may decide to leave, a temporary resident may decide to stay permanently if given the opportunity to do so. Domicile-based citizenship should therefore apply to all residents, independent of their intentions.

One could still argue that being present in a territory is not the same as residing in it. Kostakopoulou suggests that, in the context of citizenship acquisition, residency should not be simply a matter of formal presence but include "the connections and bonds of association that one establishes by living and participating in the life and work of the community" (Kostakopoulou 2008, 115). Similarly, Joseph Carens (2010) does not seem to think that mere presence should be the only eligibility criteria for citizenship. Rather, he recently argued in the context of illegalized migrants that the length of time a person is present in a political territory should define whether a migrant can acquire citizenship. Linda Bosniak, however, refutes Carens' argument by pointing to liberal constitutional norms. Based on such norms, "all persons within the state's jurisdiction are to be accorded fundamental rights, security, and recognition. For purposes of this commitment, length of stay is irrelevant; what counts is being territorially present and subject to law" (Bosniak 2010, 90). Following Bosniak, domicile – for the purpose of granting citizenship – is a matter of territorial presence rather than connections and bonds of association, length of stay, or any other arbitrary criteria.

Another practical question is: what citizenship should children have? Children typically do not voluntarily choose a place of residence but usually acquire residency through birth or through migration decisions made by their parents or guardians. One answer to this question would be to combine *jus domicilii* acquired through choice with *jus soli* acquired at birth (Bauböck 1994, 34). Kostakopoulou proposed a tripartite typology of domicile-based citizenship: first, domicile-based citizenship at birth would provide newborn children with citizenship and therefore prevent statelessness; second, domicile-based citizenship of choice would be extended on the basis of an adult's chosen residence; and third, domicile-based citizenship of association would apply to persons who are legally dependent on a citizen, such as children. A person would only be able to possess one of these types of citizenship at any given time (Kostakopoulou, 2008, 119–22). For example, when children become adults, their citizenship would transfer from domicile at birth or domicile of association to domicile of choice.

Furthermore, a question arises from the increasing global mobility of people. If they retained the domicile-based citizenships of all places in which they ever resided, they would accumulate multiple citizenships. This situation would be paradoxical because citizenship would no longer be associated with residence in a country. In fact, maintaining the citizenship of a country in which one no longer resides defies the very logic of the domicile principle. How can this paradox be solved? Rainer Bauböck (1994, 49) provides a simple solution: "If citizenship rested completely on a principle of residence a state might be entitled or even obliged to denaturalize anybody who has left the country for good." Kostakopoulou (2008, 127) agrees: "A change of residence must be accompanied by the termination of an intention to reside in the country indefinitely" and would therefore result in a loss of domicile-based citizenship.

That states revoke a person's citizenship is not unheard of. In feudal Europe, where bondage was defined in territorial terms, this bondage typically ended when a subject moved away from the territory. Seventeenth-century scholars, such as Thomas Hobbes and Ulrik Huber, affirmed this practice (Grawert 1973, 90–102). War deserters and persons committing treason could also be expelled and denaturalized. However, even the 1913 Nationality Law of the German Reich revoked citizenship from war deserters only if they were no longer residents in one of the Reich's member states (Reichs- und Staatsangehörigkeitsgesetz 1913, article 26). In this case, denaturalization occurred only under the condition that a person was a war deserter *and* not domiciled. Today, states rarely denaturalize their citizens, and usually not because somebody has moved away. The government of the USA, for example, can denaturalize citizens who join a terrorist organization or other subversive groups, provided that these citizens acquired US citizenship through immigration rather than birth. Canada recently enacted a similar law, enabling the government to revoke citizenship from naturalized citizens who possess another citizenship and who are convicted of terrorism, treason, spying, or who served in an army or armed group that engaged in armed conflict against Canada. Denaturalization solely on the basis of whether a person moved away is rarely considered a policy option.

An obstacle to denaturalizing citizens who change their country of residence is that migrants could become stateless if the country of destination does not follow the domicile principle of citizenship. This situation would put migrants at high risk. Stateless persons generally lack the right to enter and stay in a particular country. Furthermore, we know from Hannah Arendt (1985 [1948]) that states also tend to protect the universal human rights of their own citizens. Stateless persons thus have no state that looks after their human rights. To prevent statelessness, emigrants should be denaturalized only once they possess another citizenship, such as that of their new country of residence.

Revoking a person's citizenship could furthermore lead to a situation in which former citizens are denied the right to return. This situation, however, would only occur if states continued to control cross-border migration. In this case, the issue could be resolved by granting special re-entry permission to former citizens, permitting emigrants to keep citizenship if they can demonstrate that they continue to have important stakes in the country (Bauböck 2008). Under the open-borders scenario, which I discussed in the preceding chapters, the issue of the right to return would be resolved automatically. In this case, former citizens would possess – like everyone else – a right to enter a country.

The scenario of combining open borders with the domicile principle warrants further discussion. If a country adopted the domicile principle of citizenship without open borders, then only people who were born in that country or who were permitted entry would be eligible to become citizens. If migrants are denied crossing the border and are therefore unable to establish their residence in that country, then the problem of birth privilege arises again. In other words, the domicile principle of citizenship under conditions of closed borders would only reproduce the birthright of people who happen to be born on the right side of the border (Chapter 3). Domicile without open borders would be a "self-undermining" citizenship practice (Bosniak 2007, 398). To fix this problem, the right to freedom of migration should precede the right to citizenship based on residence.

The logic of this argument can also be reversed: open borders would be ineffective without domicile citizenship. If migrants were able to freely cross borders but thereafter were not entitled to citizenship, then they could still be treated unequally and be exploited as vulnerable workers, and the state could withhold protection from oppression. Although open borders and *jus domicilii* relate to different theoretical principles – the former is related to universal rights and mobility, the latter to democracy and membership – they necessitate each other.

A final looming question relates to the potential drain on national welfare systems if migrants gained access to citizenship through domicile: how could such a drain be prevented? In Chapter 4, I discussed this concern in the context of the contradictions created by the contingent possibility of open borders. Domicile-based citizenship could entitle newcomers access to the collective resources of national welfare systems. In this context, the economist Milton Friedman explained that open borders and the welfare state are only compatible if migrants are refused access to the welfare system by denying them citizenship and legal residence (Friedman 2009). The choice, according to Friedman's logic, is to either exclude migrants from status or citizenship, or to get rid of national welfare systems altogether.

Foreign residents, however, are already granted important rights in the countries where they live, and they are often rewarded for the contributions they are making to the societies in which they reside. Sociologist Yasemin Nuhoğlu Soysal (1994) found that the obligation to uphold international human rights law has driven European countries to extend domicile-based rights to foreign residents. Foreigners accumulate legal entitlements through residence and participation in civic life; they gain access to national social programs and the welfare system through paying their taxes and membership contributions. In this way, non-citizens possess "post-national" membership that in many cases makes it unnecessary to acquire formal citizenship. The downsides are that residents accumulate post-national membership gradually and on a probationary basis; that they can still lose their rights to stay and be deported; and that illegalized migrants can be denied post-national membership.

The concern – that domicile-based citizenship for all residents could drain national welfare systems – could be addressed in the following ways: first, we abolish national welfare systems altogether. Then, according to Friedman (2009), "everyone would benefit" from the presence of immigrants. Although this option is a "contingent possibility," I think it would have disastrous consequences for large sections of the population. From a social justice viewpoint, this option is unacceptable. Yet, we must be mindful that arguing for open borders and domicile citizenship might bring about the unintended consequence of the radical reduction if not outright abolition of welfare systems.

A second option is to allow migrants to acquire post-national entitlements for social programs and welfare services. In many countries, even citizens are only eligible to receive social and welfare services after they have made prior contributions. For example, they may only receive unemployment benefits after having paid into the unemployment insurance system while they were employed. The same usually applies to retirement and other benefits. Migrants with domicile-based citizenship would be treated in the same way. Basic subsistence assistance would likely not wreak havoc on public finances, since migrants tend to seek economic opportunity and a better life rather than one at the bare subsistence level. Although even eligibility for basic subsistence assistance could also be tied to residency period or contributions, any such cost-benefit calculations should take into consideration that migrants typically represent a net gain for the receiving country. A newly arriving young adult migrant, for example, "saves" a country a large amount of costs for health and social services – ranging from pre-natal care to education – accrued during childhood and adolescence before this person pays any taxes or makes any monetary contributions. In addition, such calculations should consider the future contributions that newly arriving migrants will be making. In any case, the point is that domicile as a principle of citizenship is associated with the right to stay and legal equality among all de-facto residents, while social and welfare services are typically granted based on particular circumstances, including need, prior contributions, and future benefits.

Conclusion

Next to domicile, there are other citizenship principles that could accommodate migrants. For example, *jus nexus* citizenship would require a "connection, rootedness, or linkage" to a national polity (Shachar 2011, 116). Advocacy groups – including conservative ones in the United States – are supporting *jus nexus* to frame the debate around the inclusion of illegalized migrants who are members of their communities (Sutherland Institute 2011). Similarly, "stakeholder" citizenship could be applied to migrants with a stake in the future of the country in which they live as well as to emigrants who maintain ties to their former country of residence (Bauböck 2008). Unlike domicile, however, *jus nexus* and stakeholder citizenships do not require that a citizen resides in the territorial polity. Furthermore, *jus nexus* and stakeholder citizenship are based on criteria – connections and interests – that are a matter of degree and interpretation, and that states can easily manipulate to exclude persons from membership. For example, governments and their policies often deny illegalized migrants participation in public and civic life, keep them from acquiring property, and force them to work in the informal economy and live underground. In this way, they prevent migrants from making connections and from becoming stakeholders. Thus, *jus nexus* and stakeholder citizenship lack an important feature of domicile-based citizenship: the inclusion of all residents, independent of criteria that can be denied to residents who lack citizenship. Instead, the domicile principle "makes territorial presence the all-or-nothing criterion" for citizenship (Shachar 2009, 179). This centrality of territorial presence is exactly the point: it includes all residents, independent of the state's efforts to exclude temporary or illegalized migrants, or other non-citizen residents. These migrants are de-facto members of society and are making economic, civic, and other worthy contributions. They should therefore be entitled to domicile-based citizenship.

Granting citizenship based on the domicile principle would address many of the problems of political exclusion, labor exploitation, and human hardship that temporary and illegalized migrants face. With citizenship, an important source of their vulnerability would be eradicated. Granted, migrants may still be in vulnerable positions, for example, due to practices of racial and other forms of discrimination, unequal access to social welfare benefits that must be accumulated over time, or the personal and financial costs of migration and settlement. However, the lack of formal citizenship would no longer be a reason for their criminalization, political and social exclusion, and economic exploitation. Combining open borders with domicile citizenship would be a giant step in humanity's pursuit of equality, justice, equity, and freedom from oppression.

Migrant-receiving societies would also benefit from extending domicile-based citizenship to migrants. An argument frequently mobilized against illegalized migrants is that they do not pay income and some other taxes. If these migrants received citizenship, then they would be required to pay taxes. In fact, many illegalized migrants would be ecstatic if they were permitted to contribute to society in this way. The reverse situation applies to transnational elites who sometimes evade taxes in the countries where they live by shifting capital and their legal status abroad where tax laws are more favorable. Domicile-based citizenship would undermine these practices. The domicile principle is responsive to people's *im*migration and *e*migration trajectories and treats mobile elites, migrant workers, refugees, and migrants seeking a better life or to reunite with family equally.

If the domicile principle were adopted globally, then migrants would always and only possess the citizenship of the jurisdiction in which they actually reside. Given the current state of affairs, the global adoption of the domicile principle is not a very likely possibility. On the one hand, the universal liberal appeal of the domicile principle may wield considerable moral weight in contemporary global politics. On the other hand, as nation states insist on their sovereignty and compete with each other for resources, they will probably continue to design migration and citizenship policies that cherry-pick desired migrants and exclude undesired ones. Likewise, states will likely continue to strategically manipulate their citizenship policies to exclude and exploit migrants, and to capitalize on expatriate and diaspora populations that reside outside the country. This "extraterritorial citizenship" practice (Ho 2011), too, runs counter to the principle of domicile. In practice, the global implementation of domicile-based citizenship is still a very distant reference point on the path towards the horizon that offers freedom of migration.

The principle of domicile raises contradictions, which indicate that far-reaching transformation lies in the realm of possibilia. In particular, domicile-based citizenship affirms territoriality. In fact, the very concept of "domicile" is inherently territorial because it emphasizes the territory in which one lives. The contradiction is that a territorial principle serves to accommodate populations that are increasingly mobile across territories and non-committal to any place of residence. Of course, in this chapter, I originally assumed that states occupy bounded territory. Indeed, in today's world the territoriality of states is not only a fact but the concept is so firmly entrenched in the dominant political imagination that it is hardly possible to conceive that political communities could be organized in other ways (Wimmer and Glick Schiller 2002). Territorial citizenship thus appears to be the only imaginable way to organize "redistributive politics" (Kostakopoulou 2008, 125). My practical argument therefore sought to provide an intermediate – or "meso-level" (Bauder and Matheis 2016) – policy tool to address the unequal treatment, social injustices, and forms of oppression experienced by migrants who cross borders and then reside in bounded political territories. The contradiction between territorial citizenship and human mobility is impossible to resolve theoretically within the framework of political territoriality. Contradiction, however, is a productive moment in the ongoing dialectic towards freedom, equality, and justice.

References

Arendt, Hannah. 1968. *Men in Dark Times.* San Diego, CA: Harvest Books.

Arendt, Hannah. 1985 [1948]. *The Origins of Totalitarianism.* Orlando, FL: Harvest.

Austin, Carly and Bauder Harald. 2012. "Jus Domicile: A Pathway to Citizenship for Temporary Foreign Workers." In *Immigration and Settlement: Challenges, Experiences, and Opportunities*, edited by Harald Bauder, 21–36. Toronto: Canadian Scholars' Press.

Bauböck, Rainer. 1994. *Transnational Citizenship: Membership and Rights in International Migration*. Northampton: Edward Elgar Publishing.

Bauböck, Rainer. 2008. "Stakeholder Citizenship: An Idea Whose Time Has Come?" Migration Policy Institute. Accessed April 23, 2012. http://www.migrationpolicy.org/transatlantic/docs/Baubock-FINAL.pdf.

Bauder, Harald. 2013. "Why We Should Use the Term Illegalized Immigrant." *RCIS Research Brief* 2013/1. Accessed January 26, 2016. http://www.ryerson.ca/rcis/publications/rcisresarchbriefs/index.html.

Bauder, Harald and Christian Matheis, eds. 2016. *Migration Policy and Practice: Interventions and Solutions*. New York: Palgrave Macmillan.

Blank, Yishai. 2007. "Spheres of Citizenship." *Theoretical Inquiries in Law* 8(2): 410–452.

Bosniak, Linda S. 2000. "Citizenship Denationalized." *Indiana Journal of Global Legal Studies* 7(2): 477–509.

Bosniak, Linda. 2007. "Being Here: Ethical Territorial Rights of Immigrants." *Theoretical Inquiries in Law* 8(2): 389–410.

Bosniak, Linda S. 2010. No title. In *Immigrants and the Right to Stay*, edited by Joseph Carens, 81–92. Cambridge, MA: MIT Press.

Carens, Joseph. 2010. "The Case for Amnesty." In *Immigrants and the Right to Stay*, edited by Joseph Carens, 1–51. Cambridge, MA: MIT Press.

Castles, Stephen and Davidson Alastair. 2000. *Citizenship and Migration: Globalization and the Politics of Belonging*. New York: Routledge.

Ceobanu, Alin M. and Escandell, Xavier. 2011. "Paths to Citizenship? Public Views on the Extension of Rights to Legal and Second Generation Immigrants in Europe." *British Journal of Sociology* 62(2): 221–240.

Cresswell, Tim. 2006. *On the Move: Mobility in the Modern Western World*. New York: Routledge.

European Commission. 2014. "Free Movement: Commission Publishes Guide on Application of 'Habitual Residence Test' for Social Security." Press release, January 13. Accessed November 9, 2015. http://europa.eu/rapid/press-release_IP-14-13_en.htm.

Faist, Thomas. 1995. "Boundaries of Welfare States: Immigrants and Social Rights on the National and Supranational Level." In *Migration and European Integration: The Dynamics of Inclusion and Exclusion*, edited by Robert Miles and Diethrich Thränhardt, 177–195. Cranbury, NY: Pinter Publishers.

Friedman, Milton. 2009. "Illegal Immigration." YouTube, December 11. Accessed October 18, 2015. https://www.youtube.com/watch?v=3eyJIbSgdSE.

Gosewinkel, Dieter. 2001. *Einbürgern und Ausschließen: Die Nationalisierung der Staatsangehörigkeit vom Deutschen Bund bis zur Bundesrepublik Deutschland*. Göttingen: Vandenhoeck and Ruprecht.

Grawert, Rolf. 1973. *Staat und Staatsangehörigkeit: Verfassungsgeschichtliche Untersuchungen zur Entstehung der Staatsangehörigkeit*. Berlin: Duncker and Humbolt.

Green, Simon. 2000. "Beyond Ethnoculturalism? German Citizenship in the New Millennium." *German Politics* 9(3): 105–124.

Hammar, Tomas. 1990. *Democracy and the Nation State: Aliens, Denizens and Citizens in a World of International Migration*. Avebury: Gower Publishing Company.

Ho, Elaine Lynn-Ee. 2011. "'Claiming' the Diaspora: Elite Mobility, Sending State Strategies and the Spatialities of Citizenship." *Progress in Human Geography* 35(6): 757–772.

Human Rights Watch. 2014. *World Report 2014: Events of 2013*. New York: Human Rights Watch. Accessed December 21, 2015. http://www.hrw.org.

IOM. 2014. *Global Migration Trends: An Overview.* Accessed December 16, 2015. http://missingmigrants.iom.int/sites/default/files/documents/Global_Migration_Trends_PDF_FinalVH_with%20References.pdf.

Isin, Engin F. 2012. *Citizens without Frontiers.* New York: Bloomsbury.

Jamal, Manal A. 2015. "The 'Tiering' of Citizenship and Residency and the 'Hierarchization' of Migrant Communities. The United Emirates in Historical Perspective." *International Migration Review* 49(3): 601–632,

Kostakopoulou, Dora. 2008. *The Future Governance of Citizenship.* New York: Cambridge University Press.

Levanon, Asaf and Noah Lewin-Epstein. 2010. "Grounds for Citizenship: Public Attitudes in Comparative Perspective." *Social Science Research* 39: 419–431.

OECD. 2015. "2014 Model Convention with Respect to Taxes on Income and on Capital." Accessed November 2, 2015. http://www.oecd.org/ctp/treaties/2014-model-tax-convention-articles.pdf.

Passel, Jeffrey S. and D'Vera Cohn. 2015. "Unauthorized Immigrant Population Stable for Half a Decade." Pew Research Centre, July 22. Accessed December 16, 2015. http://www.pewresearch.org/fact-tank/2015/07/22/unauthorized-immigrant-population-stable-for-half-a-decade/.

Raijman, Rebeca, Eldad Davidov, Peter Schmidt, and Oshrat Hochman. 2008. "What Does a Nation Owe Non-Citizens? National Attachments, Perception of Threat and Attitudes towards Granting Citizenship Rights in a Comparative Perspective." *International Journal of Comparative Sociology* 49(2–3): 195–220.

Reichs- und Staatsangehörigkeitsgesetz. 1913. DocumentArchiv.de. Accessed February 1, 2016. http://www.documentarchiv.de/ksr/1913/reichs-staatsangehoerigkeitsgesetz.html.

Rousseau, Jean-Jacques. 2003 [1762]. *On the Social Contract*, translated by George D. H. Cole. Mineola, NY: Dover Publications.

Shachar, Ayelet. 2009. *The Birthright Lottery: Citizenship and Global Inequality.* Cambridge, MA: Harvard University Press.

Shachar, Ayelet. 2011. "Earned Citizenship: Property Lessons for Immigration Reform." *Yale Journal of Law and the Humanities* 23: 110–115.

Soysal, Yasemin N. 1994. *Limits of Citizenship: Migrants and Postnational Membership in Europe.* Chicago, IL: University of Chicago Press.

Stevens, Jacqueline. 2010. *States without Nations: Citizenship for Mortals.* New York: Columbia University Press.

Sutherland Institute. 2011. *Immigration and Utah's Latin Problem.* Accessed April 23, 2012. http://www.sutherlandinstitute.org/uploaded_files/sdmc/Utahs%20Latin%20Problem-012011–230pm.pdf.

US Department of Homeland Security. 2015. *Profiles on Naturalized Citizens.* Accessed December 21, 2015. http://www.dhs.gov/profiles-naturalized-citizens-2013-country.

Wimmer, Andreas and Nina Glick Schiller. 2002. "Methodological Nationalism and Beyond: Nation-State Building, Migration and the Social Sciences." *Global Networks* 2(4): 301–334.

6 Sanctuary City

City air makes you free (*Stadtluft macht frei*)

Medieval legal saying

The air in Europe's medieval cities stank! It smelled of the rotting waste that residents dumped in public places, and of the manure from the pigs that roamed the muddy streets. The filth bred diseases and spread epidemics, such as the plague that killed thousands. As loathsome as the city air was, it had one redeeming quality: those who breathed it could become free. Serfs were able to shed their feudal bonds by moving from the countryside to these cities. After breathing "city air" for a year and a day, they could obtain freedom and become citizens (Schwarz 2008).

Until the beginning of the 12th century, there were only few cities in the German empire, many of which were established a millennium earlier by the Romans. The founding of Freiburg im Breisgau, which received legal autonomy in 1120 (Figure 6.1), marked the beginning of a new era. It triggered a boom in the founding of free cities throughout Central Europe. The residents of these cities were often free citizens.

This medieval legal practice of granting freedom to city dwellers contributed to the rapid growth of cities like Frankfurt am Main, Hamburg, Luneburg, and Zurich. These cities welcomed far more than 100 new citizens annually (Schwarz 2008, 108). Since medieval cities in Central Europe were small by today's standards – rarely exceeding a population of 10,000 – these numbers amounted to a considerable growth rate. The cities' capacity to accommodate their new residents must have been stretched. Nevertheless, the cities invited the migrants to stay and become citizens. As a result, the cities grew in size, accumulated wealth, and gained in political influence.

As in the past, people today flock to cities to take refuge, seek freedom, and pursue opportunity. The International Organization for Migration (IOM 2015, 39) reports that major cities, such as London, Los Angeles, New York, Melbourne, Sidney, Auckland, and Singapore, all have a foreign-born population of between 35 and 40 percent. Toronto has a foreign-born population of 46 percent, Brussels of 62 percent, and Dubai a staggering 83 percent. However, many migrants in these cities are not free citizens. In fact, those who

Figure 6.1 A tour guide dressed in medieval garb in front of Freiburg's Historical Merchant Hall
Source: Harald Bauder, 2015

crossed an international border without the national government's authorization before they settled in a particular city are often denied legal status. These illegalized migrants are deprived of equal participation in their urban communities and of fair treatment in the labor market. I am not suggesting that cities in medieval Europe were more equitable, just, or inclusive than cities today; in fact, the Jewish community was gravely discriminated against and often expelled from the city. Similarly, access to public office was often restricted to men from privileged families and denied to most ordinary citizens. It is interesting, however, that in medieval times migrants were included on an urban scale and that a corresponding legal framework – "city air makes you free" – existed. In this chapter, I explore from a contemporary vantage point the possibility that, once again, cities could become the political unit to which migrants can belong.

In the preceding chapter, I discussed how migrants could become members of the territorial nation state. Now, I explore the question of belonging in relation to urban society. I will again evoke the contingently possible, but this time I will look at the urban rather than the national scale. The work of the urban geographer David Harvey illustrates the relevance of the contingently possible for progressive urban change. Although he did not use the term "contingently possible," be observed that "the materialization of anything requires, at least for a time, closure around a particular set of institutional arrangements and a particular spatial form" (Harvey 2000, 188). Following Harvey, I examine how the city serves as the institutional arrangement and the spatial form for including migrants. Harvey also realized that such a possibility involves "tangible transformations of the raw materials given to us in our present state" (Harvey 2000, 191). He used Karl Marx's famous analogy of bees and architects to illustrate his point: architects, similar to bees, construct dwellings based on available resources and materials, space and time limitations, and existing means and techniques. Yet, unlike bees, architects also use their imagination and creativity to construct something new and potentially transformative (Harvey 2000, 199–212). In a similar way, urban political actors are creatively using the existing municipal administrations, legal frameworks, and political concepts to produce the conditions that include migrants in the urban community from which these migrants were formerly excluded. Unlike the bees who follow an inborn behavioral structure, they do not obey a pre-written script of how urban politics are supposed to occur and who belongs and who does not. By using existing structures and concepts in this subversive way, urban political actors are enacting a contingent possibility.

Utopian possibility and the figure of the city "have long been intertwined" (Harvey 2000, 156). The urban utopias of the 19th and 20th centuries envisioned by Le Corbusier, Ebenezer Howard, and Frank Lloyd Wright come to mind. These concrete urban utopias resemble the dialectical negation of the cities of their time (Fishman 1982). Like all concrete utopias, however, they aim for progressive change while harboring the potential to turn urban life into a nightmare. Harvey observed that these modern urban utopias portray the city as spaces of control, surveillance, and authoritarianism. These characteristics stifle spontancity and deny the very possibility of an open, non-scripted future. Harvey even critiqued urban hero Jane Jacobs' vision as containing "its own authoritarianism hidden within the organic notion of neighborhood and community as a basis for social life," paving the way for gated communities, shopping malls, and other structures of exclusion (Harvey 2000, 164). Along the same lines, philosopher Henri Lefebvre alleges that the 19th-century urban utopias are cleansed of politics and class struggle, and thereby deny the political nature of the urban. These depoliticized utopias depict "a city made up not of townspeople, but of free citizens, free from the division of labor, social classes and class struggles, making up a community, freely associated for the management of this community" (Lefebvre 1996, 97). What is needed today, following Harvey and Lefebvre, are not visions of a city without politics, but rather the possibilia of a future that is open and inclusive of all residents. I will turn my sight to this open future in the next chapter. Before I do this, however, let me explore the contingent possibility that the urban scale offers.

On Scale

The political geographer John Agnew (1994) has called the inability to imagine political organization in ways other than the territorial nation state the "territorial trap." This trap certainly applies to migration controls. Legal scholar Catherine Dauvergne (2008, 173) observes that "the sovereign state controlling its borders is such a powerful image that it prevents us from imagining a different way of organizing the regulation of global migration." A parallel can again be drawn between earlier periods of European history and today. Similar to the medieval feudal authority that protected serfs from enemy forces, today's nation state impersonates the sovereign protector of its citizens from outside harm, including the threat emanating from migrants who penetrate state borders. This imagination of the sovereign nation state, however, is not something that is naturally given. Political scientist Mark Salter (2011, 66) explains that sovereignty

> has no essence, and must continually be articulated and rearticulated in terms of "stylized repetition of acts" of sovereignty. The state, through its policies, actions, and customs, thus performs itself as sovereign – and this is particularly visible at borders when the self-evidence of the state's control over populations, territory, political economy, belonging, and culture is so clearly in question.

Just as we should question the essence of national sovereignty, we should also not assume that the national is the only scale at which we can frame migration and political belonging.

Yet, border and migration scholarship often reifies the national "container" when it uses this container as a taken-for-granted unit of analysis. For example, researchers tend to compile national-scale data of how many people cross an international border, how many immigrants are residing in a country, or how many were naturalized by a national government. When they analyze these data, the results are of course also organized into national categories. In this way, studies tend to reproduce the national as the only scale at which migration and belonging are imagined. Andreas Wimmer and Nina Glick Schiller (2002) call this reification of the national scale "methodological nationalism." To gain a more accurate picture of the migration process, they propose "to push aside the blinders of methodological nationalism" (Wimmer and Glick Schiller 2002, 326).

The critique of methodological nationalism also applies to the imagination of migrants belonging in a community. Why should belonging be framed at national, rather than regional, local, or other scales? Residency is often associated with geographical scales other than the nation. For example, a person may identify more with being a resident of a neighborhood, a city, or a supranational entity like Europe than a nation state. Likewise, domicile-based citizenship should not be imagined as necessarily fixed at the national scale. In fact, the "national" may not be the most intuitive scale at which the principle of domicile should be enacted. After all, the domicile principle opposes the idea that belonging is a matter of inherited membership, which is how a nation is often imagined (Chapter 5). Certainly, the current convention of privileging the national scale over other scales in regulating migration and framing citizenship reflects historical developments, but it doesn't have to be accepted as the only scale at which formal belonging is possible.

A counter position to the territorial trap and methodological nationalism is articulated by international relations scholar Stephen Krasner (2000, 124), who observes that the Westphalian model of sovereign statehood "has not constrained the imagination." Rather, he argues, political leaders have continuously created systems of authority that circumvent this Westphalian order, including empires and commonwealths (such as the British Commonwealth), supernational entities (such as the European Union), shared territory (such as Antarctica), and principalities (such as Andorra). In this way, political practice has always challenged the construct of the territorial nation state. On the one hand, Krasner's position can be used to justify military intervention and political interference in other states' affairs. On the other hand, it also permits escaping the territorial trap when contemplating alternative formations of political community and belonging.

When it comes to migration and belonging, political practices at the urban scale have presented major challenges to the construct of the territorial nation state. There is a dialectical contradiction between urban and national scales: the national scale frames the very concept of the "migrant" as a person crossing a country's border; the nation state selects which migrants are permitted to cross the border and under what circumstances; and legal practices occurring at the national scale "make people illegal" (Dauvergne 2008). Yet, it is the city where most migrants live, work, send their children to school, grow old, and eventually die. Consequently, cities confront the practical issue of including migrants in their polities and provide education, public safety, and health and other social services. For this reason, city governments and administrations are increasingly opposing the migration policies of their superior national governments. As if they were looking back at medieval Europe, cities are asserting their independence (Barber 2013).

The enforcement of national migration policies, too, occurs less and less at the national border and increasingly in cities – at workplaces, bus stations, in schools, and in public places, where illegalized migrants, for example, are identified and subsequently detained and deported. In the United States the devolution of migration-policing responsibilities has granted cities "newfound powers to discriminate on the basis of alienage, or noncitizen status" (Varsanyi 2007, 877). As I have emphasized in the preceding chapters, the possibility of progress also encompasses its dialectical negation. In this way, the urban scale can strengthen or cripple the inclusion of migrants. Being mindful of this risk, I will explore in the next section the possibility of urban belonging.

Urban Citizenship

The words "city" and "citizenship" have the same etymological origin. In fact, citizenship was originally an urban idea. Only in modern times has citizenship become associated with nationality (Sassen 2008). Today, citizenship usually expresses formal membership in the territorial nation state. In this section, I explore whether we can return to the idea of formal citizenship as an urban concept, and what that might look like in a contemporary context.

Lefebvre's (1996) work on the "right to the city" provides a useful starting point to theorize urban belonging. Lefebvre links the right to the city to *presence* in the city. Thus, people possess this right independent of being a national citizen, temporary resident, or illegalized migrant. In the words of urban geographer Mark Purcell (2013, 142): "it is the everyday experience of inhabiting the city that entitles one to a right to the city, rather than one's nation-state citizenship." Thus, the right to the city evokes the domicile principle of belonging. Everyone residing in the city has a right to the city. Lefebvre's notion of the right to the city is complex and cannot be reduced to a principle of formal membership. I will deal with Lefebvre's ideas in a more elaborate way in the next chapter. For now, I will discuss *formal* membership and apply the domicile principle to the urban context.

In Western liberal democracies, the domicile principle already applies to cities and regional polities (which in different countries are called states, provinces, cantons, or territories). In fact, "provinces and municipalities have only a single rule of automatic *jus domicili*" (Bauböck 2003, 150). They permit free entry to national citizens and accept them as full members in their communities. Local residence effectively amounts to local citizenship, including the right to vote in local elections. The problem is that local citizenship presupposes national citizenship. Foreign nationals are often not entitled to vote in local elections. Migrants who are illegalized by the national state can also expect to be excluded from formal membership in urban polities. Although migrants may live and work in cities, have children in local schools and friends in the local community, and contribute to local society, the nation state denies them local membership. It makes little sense that a *national* authority decides who is and who isn't a formal member in a *local* community.

One possible solution to this problem would be to disentangle national and urban citizenships. In this case, domicile-based urban citizenship would not require national citizenship. In the eyes of social and political theorist Rainer Bauböck (2003, 150),

> restricting urban citizenship to nationals of the state is unjustifiable whether it is imposed by national constitutions or is adopted by the local government itself. Cities should fully emancipate themselves from the rules of membership that apply to the larger state.

Bauböck proposes to strengthen the autonomy of cities and their surrounding hinterland. These city-regions could then have the authority to confer citizenship to all their residents. A family of German national citizens, for example, could be citizens of the Greater Toronto Area without assuming Canadian national citizenship. This solution, however, may create conflicts of authority assigned to urban and national scales of citizenship. For example, if the Canadian state continued to control migration over its international border, then it could deny the family of German nationals the right to return to the Greater Toronto Area after this family visited relatives in Europe or went on vacation in the USA. Interestingly, some countries are heading in the direction of implementing local domicile citizenship. In Belgium, Denmark, Finland, Ireland, Lithuania, Luxembourg, The Netherlands, Slovenia, and Sweden, residents with non-European citizenships are permitted to vote in municipal elections, provided that they have the legal permission from the nation state to be in the country. The next step would be to grant local residents who are illegalized at the national scale the right to vote in municipal elections and extend other urban citizenship rights to them. In this way, vulnerable and illegalized migrants would receive "the opportunity to both develop and express a sense of belonging that is denied to them at the national scale" (Allon 2013, 254).

Another solution would be to dispense with national birthright citizenship altogether and elevate domicile as the only citizenship principle for both national and local scales. In that case, domicile in New York or Los Angeles would coincide with domicile in the United States. A citizen of a city would always have the corresponding citizenship of the superior nation state. A solution suggested by Purcell (2002) would have the same outcome: the urban could be elevated as the scale of formal political membership over the national scale. For example, if a city grants formal urban citizenship to a resident, then this urban citizenship would automatically entitle a person to national citizenship. In that case, national citizenship practices and laws would no longer prohibit cities from granting local citizenship based on the domicile principle.

These proposals for urban citizenship based on the domicile principle present "contingent" possibilities, because we possess the conceptual tools – in particular the ideas of territorial governance and territorial citizenship – to envision such alternatives. In addition, the required municipal political systems already exist. Furthermore, formal membership in a local municipality already follows domicile rules. While currently this local membership rule applies only to national citizens, extending this rule to everyone, including illegalized migrants, is not beyond the realm of imagination. These thought experiments of envisioning the possible may inspire urban practice; ultimately, however, it is practice that generates structural change.

Urban Practice

Current political practice, to a degree, exercises domicile-based urban citizenship. In this respect, political geographer Monica Varsanyi (2007) has emphasized the importance of identity cards, so-called *matrículas consulares*, issued by the Mexican government to nationals who are living abroad. After federal US agencies tightened security in the wake of the 2001 terrorist attacks in New York and Washington, these cards became an important means of identification for illegalized Mexican migrants living in the USA. By 2005, hundreds of city administrations and more than 1,000 police agencies in the United States accepted the *matrícula consular* as valid identification. In this way, local administrations are able to acknowledge that illegalized migrants are de-facto members in their communities and deliver the services that these members need. Varsanyi (2007, 312) remarks:

> After all, formal membership in a city's polity, or "urban citizenship," is established under *jus domicili* standards in the US … [T]here are no immigration policies governing who can move into a city. City officials cannot decide who they will admit for residence and membership in their jurisdiction, and as such, formal membership in the local community (which, for instance, gives citizens the right to vote in local elections) is simply a de facto designation. These are *jus domicili* standards: if you live in the city, you're a citizen of that city.

By accepting the *matrícula consular*, cities implement the domicile principle at the local scale and in this way include migrants in the urban community.

Not only city administrations are agents of change; citizens and migrants are too. In fact, they are often the driving force behind progressive municipal policies. There are several factors that make cities important strategic locations for transformative activism: first, many migrants gravitate to the bright lights of the city, where they seek opportunity, encounter sympathetic communities, and, if necessary, can live in anonymity. Cities are therefore the places where most migrants reside, and where they form alliances with political supporters and other socially and politically disadvantaged groups. Second, migrants and citizens tend to assert their claims to political inclusion at the local scale. This scale matters most because daily life takes place locally: children go to local schools, work and consumption take place locally, and community is often framed in local terms. Third, cities are convergence points of global capital and information flows. When urban protests disrupt not only local but also national and even global economies, they amplify the political pressure they exert (Sassen 2011). As a result of these factors, cities function "as an important site of political action and revolt" (Harvey 2012, 117–18).

The 2006 migrant protests that occurred in many US cities exemplify how activists are using the urban scale to advocate the inclusion of illegalized migrants. The protests erupted in response to a proposed federal law, the so-called Sensenbrenner Bill (formally known as Border Protection, Anti-Terrorism and Illegal Immigration Control Act – HR4437), which would have made it a felony to *be* an illegalized migrant or to provide humanitarian assistance to illegalized migrants. The proposed law also contained other repressive measures against unauthorized migration. According to media reports, up to 100,000 people took to the streets in Chicago on March 10, 2006, and up to 500,000 people protested later in the spring of that year in Los Angeles and Dallas. Rallies also took place in more than 100 other US cities. Despite the seeming spontaneity of the protests, the organization required an expansive, urban, grassroots network for immigrants' rights, which took years to establish and included faith-based communities, student groups, ethnic associations, no-border activists, and ethnic media (Loyd and Burridge 2007; Pantoja et al. 2008).

Harvey cites these protests as a powerful reminder of the "collective potentiality" that rests in urban protest (Harvey 2012, 118). He also observes that the protests were "basically about claiming rights not about revolution" (Harvey 2012, 120). Rather than evoking the possibilia of radical change, the protesters opposed the criminalization of migrants by the federal government, pursuing the contingent possibility of illegalized migrants obtaining status and citizenship in the existing national community. To underscore this pursuit, the protestors evoked national symbolisms. As they walked through city streets and gathered in public squares, they sang the US national anthem, waved American (as well as Mexican, Guatemalan, and other Latin American) national flags, and proclaimed themselves to be Americans (Figure 6.2); many of the speeches held during the protests ended with "God bless America."

While these protests effectively mobilized cities as strategic locations, they did not frame citizenship and belonging in new terms or at local scales. Rather, the protesters validated the very concept of the nation and reaffirmed

Figure 6.2 Protests against Bill HR4437, Chicago, 2006
Source: Harald Bauder

the USA as the legitimate polity to which they demanded to belong. They did not seek the demotion of national citizenship from which they are excluded – nor did they demand the abolition of the American state (or the nation state in general), even though this state defines them as "aliens," enacts their disenfranchisement, and is thus the very source of their unequal treatment, exploitation, and oppression. By embracing American national symbols, the protestors pursued a political strategy of conciliation and appeasement. Their goal was to appeal to national law makers to prevent restrictive legislation from being passed, although some protesters no doubt also hoped that American law makers would take steps towards giving national citizenship to illegalized migrants. Eventually, the Sensenbrenner Bill was stopped. However, domicile-based national citizenship for illegalized migrants was not achieved. Rethinking the geographical scale of belonging never seemed to have been on the table.

Other urban activist initiatives, however, have challenged national-scale migration and citizenship policies, and successfully evoked the urban as the scale for migrant inclusion. An example is the sanctuary city movement. This movement has drawn inspiration from the sanctuary practices reported in religious scripture and applied by faith-based communities, but expanded these practices to secular urban municipalities. Initially, sanctuary activism in the USA supported military draftees who refused to fight in Vietnam and refugees fleeing from civil war in Central America. Today, sanctuary activism is most well known for campaigning in favor of municipal policies that accommodate illegalized migrants (Lippert and Rehaag 2013; Ridgley 2008).

The sanctuary city movement has had a considerable impact in numerous cities across the USA. Baltimore, Chicago, San Francisco, and dozens of other cities have passed sanctuary legislation or implemented sanctuary policies. Sanctuary measures typically ban the use of municipal resources to enforce federal immigration-related laws. For example, they can prohibit city employees from collecting and disseminating information on a person's status, and ensure the delivery of municipal services independent of a local resident's status or citizenship. With their focus on "presence" in a city (Squire 2011, 290), sanctuary policies effectively implement the domicile principle at the municipal scale. In most cases, sanctuary policies have been the result of tireless activist campaigns that lobbied municipal governments and pressured local administrations to commit to creating an environment of hospitality for migrants and refugees.

From the USA, the sanctuary city movement spread across the border to Canada. In 2004, Toronto-based activists launched a Don't Ask Don't Tell (DADT) campaign. DADT policies have been adopted by cities throughout North America as an effective tool to protect illegalized migrants. Under these policies, a city's administrative staff, municipal service providers, school boards and educators, and sometimes the municipal police force will not ask residents about their status and, if they happen to find out, will not share this information with federal authorities. These policies aim to provide illegalized residents and their children with better access to public services, libraries, education, health care, social housing, shelters, food banks, and public safety. After Toronto's city administration quietly adopted the activist demands for a DADT policy, the Toronto District School Board followed suit in 2006 and also endorsed this policy, in this way affirming all students' right to education (Berinstein et al. 2006; McDonald 2012).

The sluggish implementation and enforcement of DADT policies, however, prompted Toronto's activist community to push the matter further. The Solidarity City Network spearheaded a campaign to formalize Toronto's commitment to sanctuary practices. This network included a range of community organizations and advocacy groups, like the Law Union of Ontario, the Ontario Coalition Against Poverty, the Toronto chapter of No One Is Illegal (NOII), and Social Planning Toronto (Figure 6.3.). This network demanded that "immigration status should not be a factor in access to services and rights" (Solidarity City Network 2013, 5). All residents of the city should be able to live without fear of abuse, detention, or deportation, possess safe access to city services, be recognized as contributing members of the urban community, and be able to participate in the city's civic life.

The continued activist engagement led to the movement's greatest success so far in Canada: on February 21, 2013, Toronto became Canada's first sanctuary city. Figure 6.4 shows the Toronto City Council meeting during which a motion was adopted that affirms the city's "commitment to ensuring access to services without fear to immigrants without full status or without full status documents" (Toronto City Council 2013). Inspired by the success in Toronto, activists have been pushing other Canadian cities to adopt similar policies. The city of Hamilton, for example, became a sanctuary city in early 2014.

Figure 6.3 Solidarity City campaign, Toronto, 2013
Source: Harald Bauder

Figure 6.4 Toronto City Council, debate of "Undocumented workers in Toronto," 2013
Source: Harald Bauder

Interestingly, the sanctuary city movement was not constrained by the international border. Granted, the national scale defines cities' legal and administrative responsibilities and their capacity to initiate their own laws and policies. It also controls who is legally able to enter the country and thus settle in a city, and who will be illegalized if they do so. Furthermore, national history, immigration policy, and national public debate of immigration shape the way local communities respond to migration. Yet, activist and political networks are acting across international borders and are connecting cities like Chicago and Toronto. In fact, the activists in Toronto have been in close contact with their counterparts in US cities for inspiration, support, and practical advice on how to run a successful campaign, what concrete policies to recommend to city council, and how to ensure that these policies are actually implemented. In this way, urban practice not only challenges national authority over defining who is and who is not a member of the urban community, but urban communities are also framing themselves as increasingly independent from the national scale and the policies and identities which nation states impose on them.

Conclusion

Like the medieval cities of Europe that challenged feudal authority, cities today are again asserting their autonomy when it comes to including migrants. Sociologist Saskia Sassen (2013, 69) thinks that urban policies and practices that challenge national laws make "it possible for us to imagine a return to urban law." Sanctuary city legislation that opposes, for example, the US Department of Homeland Security's effort to involve municipal police forces in national migration enforcement may be a harbinger of such a return to urban law. However, not all cities are responding to migration in the same way. In medieval times, "city air was not the same all over" (Schwarz 2008, 11, my translation). While many cities freed serfs from feudal bonds when they became long-term residents, some cities did not follow this practice and denied serfs residency or access to citizenship. Similarly, local responses to illegalized migration today vary not only between national legal frameworks but also by factors such as a city's size, proportion of immigrant population, local labor market conditions, and local politics (Walker and Leitner 2011).

Sanctuary city policies are only one response among a wide range of options. At the other end of the spectrum, cities are enacting policies that reinforce restrictive national migration laws and discourses, some of which are even more restrictive than national policies (Gilbert 2009). A new-found autonomy of cities can produce responses that go either way – towards the inclusion or the exclusion of migrants.

In addition, we should not assume that these urban practices towards migration necessarily lead to the elevation of the urban scale over the national scale of governance. In fact, cities routinely enact national migration policies. By instating DADT and sanctuary policies, Toronto may rhetorically include all its residents, but federal "program and funding guidelines ensure that national … citizenship and the surveillance and policing of non-citizen subjects are reproduced at local levels" (Bhuyan and Smith-Carrier 2012, 205). Under current rules, national laws still outrank local ones.

The 2006 protests in Chicago and other major US cities illustrate how urban protests against the criminalization of migrants had the paradoxical effect of affirming the national scale at which migrants are criminalized in the first place. Pro-migration campaigns, such as local campaigns against the deportation of migrants, can also reaffirm ideas of national belonging (Anderson et al. 2011, 559). In contrast to the 2006 urban protests in the USA and some local anti-deportation campaigns, the sanctuary city movement has strategically shifted scale: while policies and laws at the national scale illegalize migrants, contravening sanctuary policies and legislation at the urban scale include these migrants based on their presence in the urban community, rather than their national status. At the practical level, sanctuary practices and policies have created a "de facto regularization program from the ground up" (Walia 2013, 116). It is clear, however, that the inclusion of migrants through sanctuary practices does not fundamentally challenge existing political configurations. In fact, in order to be politically feasible, sanctuary city campaigns tend to "reproduce state discourses and practices" (Czajka 2013, 54). Sanctuary city policies may undermine national immigration laws and practices, but they do not transform the structures that enable these laws and policies. Moreover, municipal sanctuary campaigns and the resulting legislation often strategically present the image of hardworking and tax-paying model migrants who are multicultural assets worthy of protection, and thereby reproduce – sometimes unwittingly but often strategically – the logic of a neoliberal economy that relies on exploitable migrant labor to begin with (Houston and Lawrence-Weilmann 2016).

Contradictions such as these motivate us to continue searching for other possibilities. Contemporary sanctuary policies and other urban practices may provide ad-hoc relief for illegalized migrants, but they are only "a small crack in the foundation of nationally defined citizenship" (Bhuyan and Smith-Carrier 2012, 217). Fundamental structural transformation is a more radical project. I will turn to this topic in the next chapter.

References

Agnew, John. 1994. "The Territorial Trap: The Geopolitical Assumptions of International Relations Theory." *Review of International Political Economy* 1(1): 53–80.

Allon, Fiona. 2013. "Litter and Monuments: Rights to the City in Berlin and Sidney." *Space and Culture* 16(3): 252–260.

Anderson, Bridget, Matthew J. Gibney, and Emanuela Paoletti. 2011. "Citizenship, Deportation and the Boundaries of Belonging." *Citizenship Studies* 15(5): 547–563.

Barber, Benjamin R. 2013. *If Mayors Ruled the World: Dysfunctional Nations, Rising Cities.* New Haven, CT: Yale University Press.

Bauböck, Rainer. 2003. "Reinventing Urban Citizenship." *Citizenship Studies* 7(2): 139–160.

Berinstein, Carolina, Jean McDonald, Peter Nyers, Cynthia Wright, and Sima Sahar Zerehi. 2006. *"Access Not Fear": Non-Status Immigrants and City Services.* Toronto. Accessed April 30, 2013. https://we.riseup.net/assets/17034/Access%20Not%20Fear%20Report%20(Feb%202006).pdf.

Bhuyan, Rupaleem and Tracy Smith-Carrier. 2012. "Constructions of Migrant Right in Canada: Is Subnational Citizenship Possible?" *Citizenship Studies* 16(2): 203–221.

Czajka, Agnes. 2013. "The Potential of Sanctuary: Acts of Sanctuary through the Lens of Camp." In *Sanctuary Practices in International Perspectives: Migration, Citizenship and Social Movements,* edited by Randy K. Lippert and Sean Rehaag, 43–56. Abingdon: Routledge.

Dauvergne, Catherine. 2008. *Making People Illegal: What Globalization Means for Migration and Law.* New York: Cambridge University Press.

Fishman, Robert. 1982. *Urban Utopias in the Twentieth Century: Ebenezer Howard, Frank Lloyd Wright, and Le Corbusier.* Cambridge, MA: MIT Press.

Gilbert, Liette. 2009. "Immigration as Local Politics: Re-Bordering Immigration and Multiculturalism through Deterrence and Incapacitation." *International Journal of Urban and Regional Research* 33(1): 26–42.

Harvey, David. 2000. *Spaces of Hope.* Berkeley, CA: University of California Press.

Harvey, David. 2012. *Rebel Cities: From the Right to the City to the Urban Revolution.* London: Verso.

Houston, Serin D. and Olivia Lawrence-Weilmann. 2016. "The Model Migrant and Multiculturalism: Analyzing Neoliberal logics in US Sanctuary Legislation." In *Migration Policy and Practice: Interventions and Solutions,* edited by Harald Bauder and Christian Matheis, 101–126. New York: Palgrave Macmillan.

IOM. 2015. *World Migration Report 2015: Migrants and Cities: New Partnerships to Manage Mobility.* Geneva: IOM.

Krasner, Stephen D. 2000. "Compromising Westphalia." In *The Global Transformations Reader: An Introduction to the Globalization Debate,* edited by David Held and Anthony McGrew, 124–135. Cambridge: Polity Press.

Lefebvre, Henri. 1996. *Writing on Cities*, translated by Eleonore Kofman and Elizabeth Lebas. Oxford: Blackwell.

Lippert, Randy K. and Sean Rehaag, eds. 2013. *Sanctuary Practices in International Perspectives: Migration, Citizenship and Social Movements.* Abingdon: Routledge.

Loyd, Jenna M. and Andrew Burridge. 2007. "La Gran Marcha: Anti-Racism and Immigrant Rights in Southern California." *ACME* 6(1): 1–35.

McDonald, Jean. 2012. "Building a Sanctuary City: Municipal Migrant Rights in the City of Toronto." In *Citizenship, Migrant Activism and the Politics of Movement*, edited by Peter Nyers and Kim Rygiel, 129–145. London: Routledge.

Pantoja, Adrian D., Cecilia Menjívar, and Lisa Magaña. 2008. "The Spring Marches of 2006: Latinos, Immigration, and Political Mobilization in the 21st Century." *American Behavioral Scientist* 52(4): 499–506.

Purcell, Mark. 2002. "Excavating Lefebvre: The Right to the City and Its Urban Politics of the Inhabitant." *GeoJournal* 58(2–3): 99–108.

Purcell, Mark. 2013. "Possible Worlds: Henri Lefebvre and the Right to the City." *Journal of Urban Affairs* 36(1): 141–154.

Ridgley, Jennifer. 2008. "Cities of Refuge: Immigration Enforcement, Police, and the Insurgent Genealogies of Citizenship in US Sanctuary Cities." *Urban Geography* 29(1): 53–77.

Salter, Mark. 2011. "Places Everyone! Studying the Performativity of the Border." *Political Geography* 30(2): 61–69.

Sassen, Saskia. 2008. *Territory, Authority, Rights: From Medieval to Global Assemblages*, updated edition. Princeton, NJ: Princeton University Press.

Sassen, Saskia. 2011. *Cities in a World Economy*, 4th edition. Thousand Oaks, CA: Pine Forge.

Sassen, Saskia. 2013. "When the Center No Longer Holds: Cities as Frontier Zones." *Cities* 34(October): 67–70.

Schwarz, Jörg. 2008. *Stadtluft macht frei: Leben in der mittelalterlichen Stadt.* Darmstadt: Primus Verlag.

Solidarity City Network. 2013. *Towards a Sanctuary City: Assessment and Recommendations on Municipal Service Provision to Undocumented Residents in Toronto.* Toronto. Accessed December 20, 2013. http://solidaritycity.net/learn/report-towards-a-sanctuary-city/.

Squire, Vicki. 2011. "From Community Cohesion to Mobile Solidarities: The City of Sanctuary Network and the Strangers into Citizens Campaign." *Political Studies* 29(2): 290–307.

Toronto City Council. 2013. "Motion CD18.5: Undocumented Workers in Toronto." Accessed February 5, 2016. http://app.toronto.ca/tmmis/viewAgendaItemHistory.do?item=2013.CD18.5.

Varsanyi, Monica W. 2007. "Documenting Undocumented Migrants: The Matrículas Consulares as Neoliberal Local Membership." *Geopolitics* 12(2): 299–319.

Walia, Harsha. 2013. *Undoing Border Imperialism*. Oakland, CA: A. K. Press.

Walker, Kyle E. and Helga Leitner. 2011. "The Variegated Landscape of Local Immigration Policies in the United States." *Urban Geography* 32(2): 156–178.

Wimmer, Andreas and Nina Glick Schiller. 2002. "Methodological Nationalism and Beyond: Nation-State Building, Migration and the Social Sciences." *Global Networks* 2(4): 301–334.

7 Right to the Future

> We strongly believe that none of us are free until all of us are free.
>
> No One Is Illegal (2013)

Sanctuary city policies accomplish more than just providing refuge to illegalized migrants. They also enable these migrants to be active members in their communities. In sanctuary cities, illegalized migrants can pick up their children from school without fear of being detained and deported; they can participate in municipal programs; and they can call the police to report a crime. No longer being forced into constant hiding, they have opportunities to participate in public life. In this way, sanctuary cities permit illegalized migrants to take part "in the everyday enactment of the city through its routines, practices and rhythms" (Darling and Squire 2013, 210). In Toronto and other sanctuary cities throughout North America, community organizers have been quite cognizant of the limited impact of their campaigns on federal immigration policy. However, the importance of their campaigns lies in their "ability to change the ways in which people interact with one another locally and to develop a shift in ideas around community and belonging" (McDonald 2012, 143). These interactions and shifts in the ideas around community and belonging are a critical ingredient for fundamental social and political transformation. They are what philosopher Henri Lefebvre had in mind with the notion of the "right to the city." This notion refers not only to the inclusion of all of the city's residents but also to the manner in which everyday practice fundamentally transforms society. The title of this chapter makes reference to this concept, albeit in a way that emphasizes future and possibility rather than urban space per se.

The concept of the sanctuary city, which I examined in the preceding chapter, presupposes that the basic structure of urban governance persists. Sanctuary cities make use of municipal administrations and city councils, which develop and pass policies and laws that apply to their territorial jurisdiction. For many migrants who live in cities without sanctuary policies, the sanctuary city is still only a contingent possibility. Other cities have already successfully implemented sanctuary policies. In this chapter, I explore possibilia, which assumes the fundamental transformation of the social and political structures that we take for granted today. Although urban geographer David Harvey does not use the terms contingent possibility and possibilia, he recognizes the difference between both levels of possibility. Urban rebellions, he writes, will have to be

> consolidated at some point at a far higher scale of generality, lest it all lapse back at the state level into parliamentary and constitutional reformism that can do little more than reconstitute neoliberalism within the interstices of continuing imperial domination. This poses more general questions not only of the state and state institutional arrangements of law, policing, and administration, but of the state system within which all states are embedded.
> (Harvey 2012, 151)

To keep all options open, we must question not only the state and its institutions, but all social and political structures and the ideas through which we understand the world.

The impossibility of envisioning possibilia in concrete terms should not discourage us from embarking on the journey towards it. This point is not lost on the activists in Chicago, Toronto, and other cities. In addition to demanding immediate policy changes, they also aspire to a world beyond exclusionary categories, such as "migrant." Their practices defy these categories, yet they refrain from painting a concrete picture of this future. Instead, their practices create alliances and forge solidarities that are critical milestones on the journey towards possibilia.

From Identity Formation …

Scholars and political activists are keenly aware that the process of identity formation lies at the heart of societal transformation. For Karl Marx and Friedrich Engels, revolution required that the urban working class (which existed as a social fact, or *in itself*) realizes that it constitutes a political force (and exists *for itself*). They saw the formation of a collective identity as the key for the working class to "lose its chains" and for humanity to overcome class-based society.

Since Marx and Engels, the structure of society has changed. In the mid-19th century, Marx and Engels had identified two antagonistic urban classes: the bourgeoisie and the proletariat. These classes no longer exist in the same way. By the 1930s, the philosopher Theodore W. Adorno observed that "the proletariat has more to lose than its chains" (cited in Hawel 2006, 112, my translation). Today, the majority of the population of the global north identifies with the "middle class." The members of this class possess social and economic rights, access to health services, a relatively high standard of living and life expectancy, the latest entertainment technology, and access to educational systems for themselves and their children. Although this "middle class" has witnessed their labor rights and welfare entitlements withering away over the last decades, society has not returned to the class structure of the Industrial Revolution.

Today's agents of social and political transformation must be theorized in different ways. Harvey suggests that the concept of class is not without redeeming qualities. Rather than challenge the concept of class itself, he proposes an understanding of class that is more inclusive than the former proletariat. In his words:

> There is no proletarian field or utopian Marxian fantasy to which we can retire. To point to the necessity and inevitability of class struggle is not to say that the way class is constituted is determined or even determinable in advance.
>
> (Harvey 2005, 202)

A few years after publishing this remark, Harvey elaborated on who this concept of class may include:

> So we now have a choice: mourn the passing of the possibility of revolution because that proletariat has disappeared or change our conception of the proletariat to include the hordes of unorganized urban producers (or the sorts that mobilized the immigrant rights marches) and explore their distinctive revolutionary capacities and powers.
>
> (Harvey 2012, 130, parentheses in original)

According to Harvey, the agents of social and political transformation today bridge the various dimensions of race, gender, sexuality, and other markers "that are closely interwoven with class identities" (Harvey 2005, 202). Lefebvre agrees that societal transformation may involve the "pressure of the working class," but the working class alone is "not sufficient" (Lefebvre 1996, 157).

Another useful way to capture the structure of today's society is through the concept of the "precariat" (Standing 2011; Harvey 2012). This concept includes the underpaid service worker who has two minimum-wage jobs to make ends meet, the casually employed single parent whose child lacks access to proper health care and education, the unemployed middle-aged manufacturing worker without the prospect for a decent pension, and the illegalized migrant who is denied basic economic and social rights. In other words, the precariat encapsulates workers who are deprived of their fair share of the value that their labor produces, persons who lack opportunity, as well as people who are denied rights and entitlements. Correspondingly, calls for societal transformation are articulated around notions of both social justice and citizenship. Or, in Harvey's (2012, 153) words: "Citizen and comrade can march together."

The situation of illegalized migrants illustrates how matters of social justice and citizenship are interlaced. The denial of formal citizenship renders illegalized migrants superexploitable as workers. In fact, illegalized migrants exemplify the contemporary precariat like few other groups: they are denied formal citizenship *and* are superexploited as labor. With a nod to the proletariat of the 19th century, philosopher Étienne Balibar (2000, 42) has called illegalized migrants the "modern proletariat." Unlike the proletariat of the 19th century, however, illegalized migrants are not a distinct class but a part of a larger social formation consisting of persons experiencing various forms of exclusion, injustice, and oppression. This formation can become a powerful force for political transformation. It lacks, however, a homogenous identity. Unlike the former proletariat that could be defined by its role in the industrial production process and its lack of ownership of the means of production, this social formation is inherently heterogeneous and defined by the connections and interdependencies among people in diverse – albeit precarious – social situations and legal circumstances. It does not possess a unifying identity that can be grasped with current ways of thinking. As part of this larger social formation, illegalized migrants can be seen as "emerging as something more, something else, something other" (Nyers 2010, 141). For this emerging social formation to unfold its transformative capacity requires solidarity.

… to Solidarity

Even though illegalized migrants and other excluded groups are denied full social and political participation, they are no less a part of society. Slaves and their masters both are proficient in a common language, otherwise the master would not be able to give commands that the slaves understand. But when the slaves begin to utter their own demands, they become political actors (Rancière 1999, 2004). A similar situation occurred during the protests in Chicago and other US cities in 2006, where illegalized migrants took to the streets and proclaimed that they, too, are Americans. Effective politics, however, also require a critical mass of voices and bodies. As the 2006 protests also showed, acts of solidarity can amplify voices and multiply bodies.

Solidarity between people who are different is not an oxymoron – quite to the contrary. From a Hegelian perspective, identity always requires the reference point of the other. "Because the subject only comes into being via the other it has a general debt towards the other" (Kelz 2015). This debt should not be understood as guilt, empathy, or self-serving utility. Rather, it is an integral part of the dialectic of subject formation. It bonds people who are in different social and political circumstances, such as citizens and illegalized migrants. In fact, alliances that span such divides can be powerful forces of social and political transformation. Political scientist Heather Johnson studied how different groups contest their political exclusion. She observed how refugees in Tanzania, unaccompanied youths in Spain, and asylum seekers in Australian detention centers all chose to transgress from the conduct and behavior expected of them. In this way, they became political actors. However, the transformative potential is enabled only with "the establishment of a relationship of solidarity between non-citizens and citizens" (Johnson 2012, 117). Granted, the relationships of solidarity between formal citizens and non-citizens are asymmetrical (Kelz

2015); however, the point is not that the citizen speaks on the behalf of the non-citizen, but that non-citizens and citizens now share a political stage on which the voice of the non-citizen is heard.

The 2006 protests in Chicago and other US cities involved similar alliances between citizens and non-citizens. These protests included, for example, non-migrants, naturalized Americans, long-established Chicano groups, and illegalized migrants. Protesting together gave illegalized migrants a shared political sphere to "make a mimetic claim to citizenship" (Butler 2012, 122). The sanctuary city movement, too, has mobilized the solidarity of formal citizens and their representatives on municipal councils. The shared political sphere created by the bond of solidarity between formal citizens and illegalized migrants has enabled the latter to "enact themselves as political subjects in their own right" (Squire and Bagelman 2012, 147).

I have shown in the previous chapter how the 2006 protests and sanctuary city movement grasp the contingent possibility of formal belonging in existing national and urban polities. However, the category of formal citizenship may not be suitable to capture the identities of "peoples whose actions may not necessarily be framed in this way" (Nyers and Rygiel 2012, 10). While formal citizenship has the potential to be inclusive, it also excludes and draws borders between people who belong and people who do not. Sanctuary cities, for example, continue to distinguish between "guest and host" (Darling and Squire 2013, 193–4). Transcending the categories of guest and host, migrant and non-migrant, citizen and non-citizen, indigenous and settler lies in the realm of possibilia.

Practice in Action

In Chapter 4, I associated possibilia with the notion of no border. Activist scholar Nandita Sharma (2013) recently said that no border is "not a political proposal – it's a revolutionary cry." She meant that a no-border world entails the fundamental reconfiguration of the way people live together and govern themselves. The conditions, practices, and ways of thinking that characterize such a world do not yet exist.

The no-border network illustrates how practice can evoke possibilia. This network consists of a coalition of groups and activists from Germany, Italy, the United Kingdom, and other European countries. In the early 2000s, this network organized no-border camps throughout Europe, in places such as Strasbourg (France), Rothenburg (Germany), near Białystok (Poland), Tarifa (Spain), and Trassanito (Italy). The purpose of these camps was "to allow refugees, migrants and undocumented migrants, such as the 'San Papiers' in France, and the members of support/campaign groups from across Europe to forge new alliances and strengthen solidarities" (Alldred 2003, 153). The Strasbourg no-border camp was in a symbolic location, not only near the French–German border, but also in the city that represents European integration, the triumph over national hostilities, and the fall of border barriers. It was attended by an estimated 2,000 to 3,000 participants and consisted of several "barrios," each arranged around a communal kitchen. Political scientist William Walters (2006, 30) argues that these camps cultivated a "milieu of solidarity and self-identity" that broke up the binaries of citizens and aliens, nationals and foreigners, or migrants and non-migrants. The no-border camp

> is not just demanding freedom of movement, but is in some small way enhancing it. The modern state defines territory by striating and monopolizing space. Rather like camping in a wilderness area, border camping seems to imply a different relationship to the land.
>
> (Walters 2006, 32–3)

Although the milieu of the no-border camp is temporary and geographically contained, it defines a political practice that gives us a glimpse of possibilia. In this way, the no-border camp conjures a utopia moment.

The no-border network also inspired an initiative called Borderhack, held in Tijuana, on the border between Mexico and the United States. The 2001 theme of Borderhack was "delete the border." It focused on "eliminating the mental blockades" that cause people on both sides of the border to take the border for granted. The organizer, Fran Ilich, said:

> I grew up thinking the border was a very normal thing … It was something I never questioned. But as I got older, I started noticing that persons from the United States would come to party, but wouldn't bother to get to know the Mexican people. They just saw us as workers, not equals.
>
> (Scheeres 2001)

"Deleting" the border erases these distinctions and opens a possibility for solidarity and new identities to emerge.

The practices of No One Is Illegal (NOII) activists are another example that illuminate a pathway towards possibilia. Originally, the phrase "no one is illegal" emerged in the 1950s in response to Operation Wetback, a law-enforcement program by the United States government targeting illegalized migrants mostly from Mexico (Anderson et al. 2009). Today, NOII is active worldwide, including Germany, where in 1997, German activists, anti-racism organizations, unions, and other groups established the network *Kein Mensch ist Illegal* (No One Is Illegal). NOII was a response to the illegalization of migrants, and it demanded rights for all residents irrespective of formal citizenship and legal status. Correspondingly, NOII activities reject the identities and categories the state imposes upon people. Much of the practical work by NOII remains hidden from state authorities and the general public to protect illegalized migrants. However, other activities make a point of being highly visible as a political strategy.

In 1999, the migrant Aamir Ageeb died in the custody of three officers of the German border protection force (*Bundesgrenzschutz*) while being deported to

Sudan aboard a Lufthansa airplane. In response to this tragedy, NOII launched the "Deportation Class" campaign against the forceful removal of illegalized migrants via popular airlines, such as Lufthansa. The campaign's name draws attention to the fact that not all passengers book their flights in first, business, or economy class to travel to business meetings or reach vacations spots; some passengers are forced to travel "deportation class" to a destination where they do not wish to go. The activities of this campaign included protesting at Lufthansa's annual shareholder meeting and disrupting boarding procedures in an attempt to prevent the departure of planes with deportees on board (Stierl 2012). These activities relied on the solidarity between formal citizens and migrants who the state has illegalized. In this case, the participation of formal citizens does not open up an equally shared political stage. Rather, the participation of citizens is necessary because citizens are able to speak and protest without suffering the same consequences as the migrants, who are in much more vulnerable positions. As NOII activists are well aware, the challenge lies in creating a shared political sphere "without recreating helper-victim power dichotomies" (Stierl 2012, 435).

NOII has also been active in Canada. NOII activities in Canada are situated in a different geographical, historical, and political context than in Germany. The practices by NOII activists in Canada, I think, exemplify how solidarity and identity formation interlock, and in this way begin leveling the uneven playing field between citizens and non-citizens, migrants and non-migrants, and so on within the shared political stage. Although NOII's core activities revolve around supporting illegalized migrants, Canadian NOII activists explicitly express solidarity with other groups and individuals who are suffering from various forms of oppression. This expression of solidarity could be observed at the Annual May Day of Action in Toronto (Figure 7.1), where

Figure 7.1 Annual May Day of Action, Toronto, 2013
Source: Photo by Harald Bauder

NOII and Solidarity City were at the forefront of a coalition of social justice groups, labor unions, community organizations, anti-poverty advocates, charities, Indigenous groups, and other organizations. They marched together against ableism, colonialism, environmental destruction, homophobia, imperialism, patriarchy, racism, and transphobia (No One Is Illegal 2012).

In particular, the alliance between migrant-supporting and Indigenous organizations illustrates the activists' critical practice. This alliance rejects the way in which the mainstream media and politics portray migrants and Indigenous peoples as antagonistic in respect to how both groups claim territorial belonging: in the one corner is Canada, the settler society, which places migration at the center of its national imagination. This Canada cannot be imagined without foreign migrants, who came, settled, and built the Canada we know today. In the other corner are Indigenous peoples, who claim territorial belonging based on the principle of ancestry. According to this construction, migrants and Indigenous peoples are dialectical opposites (Bauder 2011; Sharma and Wright 2008–9). The solidarity between NOII and Indigenous organizations rejects those antagonistic constructions. By acting in unity, the activists rebuff the categories that divide migrants and Indigenous peoples, citizens, and non-citizens. Instead, they affirm their shared experiences of racialization and dispossession, and their common struggle against the distinction, exploitation, and oppression that borders have created (see Chapter 2). The resulting liberatory vision "is one that is based less on pathways to citizenship in a settler state, than on questioning the logics of the settler state itself" (Walia 2013, xiii). Activist Ruby Smith Días of the Vancouver chapter of NOII eloquently links the idea of Indigenous sovereignty with freedom of migration:

> For me, the notion of free migration and Indigenous sovereignty are not contradictory. People have always moved – whether for food, safety, celebration, love. What matters in most cases was that respect for the land and peoples in that area would be upheld. That we don't see our struggles as separate, but as relationships of solidarity. So let's dream on. Let's build our dreams together.
>
> (Walia 2013, 237)

Smith Días' dream resonates strongly, I think, with the possibilia that lies at the utopian horizon, but that cannot yet be grasped in concrete terms.

The solidarity on the side of migrant activists is reciprocated by Indigenous activities that are supporting migrant struggles. In 2010, Indigenous activists in Arizona organized against the criminalization and oppression of illegalized migrants called for in the state legislation titled Support Our Law Enforcement and Safe Neighborhoods Act (or SB1070). The Indigenous activists made the following demands:

> Indigenous communities such as the O'odham, the Pascua Yaqui, Laipan Apache, Kickapoo, and Cocopah along the US/Mexico border have been

> terrorized with laws and practices like SB1070 for decades ... Many people are not able to journey to sacred sites because the communities where people live are on the opposite side of the border from these sites. Since the creation of the current U.S./Mexico border, 45 O'odham villages on or near the border have been completely depopulated.
>
> On this day people who are indigenous to Arizona join with migrants who are indigenous to other parts of the Western Hemisphere in demanding a return to [the] traditional indigenous value of freedom of movement for all people.
>
> (O'odham Newswire 2010)

In Canada, Indigenous elders welcomed 492 Tamil refugees who arrived aboard a ship named the *MV Sun Sea* in 2010 on the Canadian West Coast. This welcome contrasted with the reception by the Canadian federal government, which detained most of the refugees.

Similar acts of solidarity have occurred in other contexts. For example, the conference "Building a Solidarity City," which took place in Montreal in 2013, sought to assure "access to free and quality services related to health, education, food, housing, shelters and more, for non-status migrants and all residents of Montreal." While this conference included diverse migration-related workshops, discussions, and panels on "Deportation, Prison & the Double Punishment of Migrants," "Immigration Support: A Strategy Session," "No Borders Movements & Building Solidarity Cities Across North America," these topics were also connected to "Indigenous Sovereignty & Self-Determination" (Cité sans frontières 2013a). The situation of migrants and Indigenous peoples was framed as a shared problem of wider structural oppression.

In December 2013, activists in Quebec mobilized against the proposed Quebec Charter of Values, which would have infringed on the rights of religious minority and migrant groups in Quebec. The activists' statement opened with an expression of solidarity with Indigenous struggles and a call to end all forms of oppression:

> From the outset, the proposed Charter and related debate fails to recognize that Quebec and Canada are built on stolen Indigenous land, and constituted through the dispossession and genocide of Indigenous peoples. We assert our solidarity and support with Indigenous struggles for self-determination and cultural integrity.
>
> We are for equality between all genders but we also assert our support for struggles against patriarchy, sexism, homophobia, transphobia, racism and all forms of oppression.

This statement was signed by a wide range of organizations, including the Centre de lutte contre l'oppression des genres/Centre for Gender Advocacy, the Immigrant Workers Center/Centre des travailleurs et travailleuses immigrantEs,

No One Is Illegal/Personne n'est illégal-Montréal, and Solidarité sans frontières/ Solidarity across Borders (Cité sans frontières 2013b).

Will these acts of solidarity result in the awakening of a shared consciousness among a diverse precariat? I think it is premature to answer this question. It may well be possible that acts of solidarity can bring about a common identity that encompasses those we describe today as illegalized migrants, Indigenous peoples, the racialized and criminalized, the disabled, displaced, deprived, and dispossessed. However, it is also possible that these acts of solidarity produce a new consciousness among a diverse social formation that we cannot yet fully grasp. The point, I think, is that the possible outcome remains open.

Scalar Practice

The imagination of utopia has always involved a particular geographical scale. Thomas More's utopia consists of an island defining the territory of a republic. Le Corbusier, Ebenezer Howard, and Frank Lloyd Wright focused on the urban scale. The utopia of H. G. Wells, with its freedom of movement, entailed a "World State speaking one common tongue" (Wells 1959 [1905], 41). The question of scale also arises in the context of the social and political practices that evoke the possibilia of an inclusive society offering freedom of migration.

When urban-economic geographer Michael Samers (2003) contemplated a borderless world, he refrained from articulating a concrete utopia. Nevertheless, he believed that the demise of migration controls at the national scale will require "compensating measures at some other scale" (p. 214). He writes:

> I am calling less for a sketch of utopia than a non-teleological imaginary of global society. If imagination is to lead to practice, then this will have to involve an individual and collective exertion over the question of a cosmopolitan justice at another scale. Such is our task ahead.
>
> (Samers 2003, 216)

The global scale, Samers suggests, is where such compensating measures could be enacted.

Bridget Anderson and her colleagues have also considered the global scale in the form of a global commons governing a borderless world (Anderson et al. 2009). Drawing on the historian Peter Linebaugh's (2008) work on the practice of communing, they argue that freedom of migration and the right to stay is a "common" right, which exists globally. However, the common right must not be understood as an abstracted right, like today's human rights. Rather, the common right is an entitlement that is always situated in particular social, political, historical, and geographical circumstances. Due to its contextualized nature, the common right cannot be grasped in concrete terms in a not-yet-existing world.

However, even as a contextualized concept, the "commons" and the right to the commons may at some level already fix what is possible. Harvey, for example, applies the idea of the commons "to keep the value produced under the control of the laborers who produced it" (Harvey 2012, 87). Adorno would likely have responded to Harvey and Anderson and her colleagues that the possible must never be drawn up based on existing ways of thinking – including the concept of the commons and the right to the commons, or the economic understanding that labor produces value – because these concepts and ways of thinking will be re-enacted in this way.

Europe, representing the supranational scale, also features prominently in imagining the political possibility of freedom of migration. Sociologist Ulrich Beck and political scientist Edgar Grande (2007) lament that the national imagination has stifled the emergence of a cosmopolitan Europe. They envision existing national borders within Europe to serve as meeting places rather than barriers, enabling people to belong on either side of the border and in solidarity with each other. Political theorist Sonja Buckel and her colleagues (2012), too, are imagining Europe as an emancipatory project. Mere scale shifting, however, only reproduces practices of exclusion – albeit at a different scale. The European Union's Schengen rule of free mobility and the creation of European citizenship have coincided with the hardening of Europe's external border. This border disproportionately targets racialized and poor migrants and people fleeing from war and violence. Today, Europe's external border is the most deadly in the world.

When it comes to transformative social and political practice, the urban emerges again as the critical scale. I have discussed this scale in the preceding chapter in the context of urban protests and sanctuary cities. Sociologist Saskia Sassen proposes that "the loss of power at the national level produces the possibility for new forms of power and politics at the sub-national level" (Sassen 2008, 314). Her ground-breaking work on cities suggests that the urban scale will fill the void left by the dwindling sovereignty of the nation state (e.g. Sassen 2011). Sassen connects the future of cities with their role in medieval Europe:

> cities can accommodate and enable the unbundling of the right articulation of the citizen and formal state politics. These various trends resonate with the case of the burghers in the medieval cities: they were informal actors who found in the space of the city the conditions for their source of "power" as merchants and for their political claim making. In my interpretation, complex cities today also function as such a productive space, but for different types of informal political actors and claim-making.
>
> (Sassen 2008, 321)

The city provides the context not only of formal belonging but also where new social formations articulate their political claims. Such claims were also made during the 19th and early 20th centuries in the cities, which were core sites of production, exploitation, and class struggle and were therefore the locations where proletarian identities and revolutionary movements were forged

(Merrifield 2002). Today, the urban continues to be a catalyst for acting in solidarity and forming new identities.

Citizenship scholar Engin Isin offers an additional perspective on the importance of the urban scale. He suggests that the city differs from other structures like the nation state. The city "exists as *both* actual and virtual spaces" (Isin 2007, 212, emphasis in original). As actual space, the city encompasses a physical infrastructure consisting of houses, streets, public spaces, and so on. This actual space of the city brings people in physical proximity to each other. It is where "bodies congregate" (Butler 2012, 117), which is a crucial condition of "being political" (Isin 2002; 2008). By facilitating the political, "cities are one of the key sites where new norms and new identities are *made*" (Sassen 2013, 69, emphasis in original). In this way, cities provide the actual space for social and political practices that evoke possibilia. As virtual space, the city is imagined in certain ways. The city, for example, possesses a structure of governance and legal institutions, but these institutions come into being only when they are enacted through the city's actual space. In contrast to the city, the nation state exists *only* as a virtual space – as an "imagined community" (Anderson 1991). The same applies to supernational entities like the European Union or ideas, such as the global village or the global commons. In fact, all these different scales of imagined communities are enacted through the city's actual space. Thus, the urban is necessarily the scale at which possibilia is enacted.

The urban is also the scale of Lefebvre's notion of the right to the city. This notion refers to urban politics that offers "a radical alternative that directly challenges and rethinks the current structure of both capitalism and liberal-democratic citizenship" (Purcell 2002, 100). Drawing on Hannah Arendt and David Harvey, philosopher Eduardo Mendieta concludes that the right to the city involves not only legal entitlements but also the "right to determine the ways in which we can define and transform ourselves" (Mendieta 2010, 445). In other words, the inhabitants of the city – including illegalized migrants – possess the freedom to begin something new that has not existed before (Arendt 1960, 32). Thus, they have the right to bring about possibilia.

Lefebvre imagines the possible city as unfixed and "experimental" (Lefebvre 1996, 151). This city is neither a concrete model, like the modernist urban utopias conceived by Le Corbusier, Ebenezer Howard, and Frank Lloyd Wright, nor a vision defined in contemporary concepts or ways of thinking. Harvey (2012 140, parentheses in original) comments on Lefebvre's ideas of urban politics in this way: "Most of what we know about urban organization comes from conventional theories and studies of urban governance and administration within the context of bureaucratic capitalist governmentality (against which Lefebvre quite rightly endlessly railed)." Instead of developing a concrete utopia, Lefebvre imagines the possible city as open.

Lefebvre's understandings of urban politics corresponds with what Bloch (1985 [1959]) called the "real" possible, and what I call possibilia. Lefebvre's term "possible-impossible" describes the same idea. Architectural scholar

Nathaniel Coleman (2013, 353, emphasis in original) explains: "Bloch's conception of the *Real-Possible* is akin to Lefebvre's *Possible-Impossible.*" Lefebvre (1996, 181) clarifies that his imagination of the city projects "on the horizon a 'possible-impossible'" that neither denies the present nor the past; yet, the inhabitants of this possible-impossible city embrace identities that differ from those that conventional thinking defines today or has defined in the past. In this city, the very concept of "migrant" may no longer be relevant and thus be used to be exclude and illegalize members of the urban community. However, we do not yet know what, if any, alternative identities these urban inhabitants will adopt.

Conclusion

The future is wide open. This means that even concepts such as the "urban," which are referenced in particular historical and geographical contexts, should not be taken for granted. In fact, geographers have long appreciated that scale is a human invention created to make sense of the world. It is probable that the future will bear scales of governance and human co-existence that we cannot imagine from our contemporary vantage point.

Along the road towards the distant possibilia lie contingent possibilities that frame the world in today's concepts and ways of understanding the world. The sanctuary city, which I described in the preceding chapter, is such a contingent possibility. The philosopher Jacques Derrida (2001) takes this idea of the sanctuary city a step further when he discusses the "city of refuge." On the one hand, he holds on to the urban scale when he pursues a "dream of a novel status for the city" (2001, 3). On the other hand, he envisions a new type of urban politics and novel "forms of solidarity yet to be invented" (4) that enact not-yet-existing "modalities of membership" (4). Derrida's city of refuge seems not quite yet comprehensible from our current vantage point.

Transformation towards the unknowable will proceed in a dialectical fashion. Acts of solidarity are interventions in this dialectic. These acts not only bridge the distinctions between migrants, non-migrants, illegalized persons, citizens, Indigenous peoples, city, and nation, but they also challenge the essence of these distinctions. In fact, acts of solidarity may produce the consciousness of new political actors that, in turn, may give rise to new ways of living together.

Radical transformation, however, also harbors the danger of dialectical negation. If the territorial nation state loses the monopoly of control over movement, access to territory could easily follow a gated-community model of private ownership (Torpey 2000, 157). The citizenship of states like Austria, Antigua and Barbuda, Cyprus, Malta, and St. Kitts and Nevis is already up for grabs in exchange for economic investments or charitable donations; other countries, like Canada, have separate immigration programs for economic elites. Similarly, it would be frightening to return to a pre-modern political order, as envisioned by the leaders of the recently established Islamic caliphate on the territory of Syria and Iraq that does not recognize state borders. The reorganization and rethinking of human migration and belonging require continual reflection and participation in the simultaneous transformation of practices and structures of migration and belonging, and the corresponding ways of understanding the world. At no point will engagement no longer be necessary.

References

Alldred, Pam. 2003. "No Borders, No Nations, No Deportations." *Feminist Review* 73: 152–157.

Anderson, Benedict. 1991. *Imagined Communities: Reflections on the Origin and Spread of Nationalism*, revised edition. London: Verso.

Anderson, Bridget, Nandita Sharma, and Cynthia Wright. 2009. "Why No Borders?" *Refuge* 26(2): 5–18. Accessed October 4, 2011. http://pi.library.yorku.ca/ojs/index.php/refuge/article/viewFile/32074/29320.

Arendt, Hannah. 1960. "Freedom and Politics: A Lecture." *Chicago Review* 14(1): 28–46.

Balibar, Étienne. 2000. "What We Owe to the San-Papiers." In *Social Insecurity*, edited by Len Guenther and Cornelius Heesters, 42–44. Toronto: Anansi.

Bauder, Harald. 2011. "Closing the Immigration-Aboriginal Parallax Gap." *Geoforum* 42(5): 517–519.

Beck, Ulrich and Edgar Grande. 2007. *Das kosmopolitische Europa: Gesellschaft und Politik in der Zweiten Moderne* (The Cosmopolitan Europe: Society and Politics in the Second Modernity). Frankfurt am Main: Suhrkamp.

Bloch, Ernst. 1985 [1959]. *Das Prinzip Hoffnung*. Frankfurt/Main: Suhrkamp.

Buckel, Sonja, Fabian Georgi, John Kannankulam, and Jens Wissel. 2012. "'… wenn das Alte nicht stirbt und das Neue nicht zur Welt kommen kann.' Kräfteverhältnisse in der europäischen Krise." In *Die EU in der Krise: Zwischen autoritärem Etatismus und europäischem Frühling*, edited by Forschungsgruppe Staatsprojekt Europa, 12–48. Münster: Westfälisches Dampfboot.

Butler, Judith. 2012. "Bodies in Alliance and the Politics of the Street." In *Sensible Politics: The Visual Culture of Nongovernmental Activism*, edited by Meg McLagan and Yates McKee, 117–137. New York: Zone Books.

Cité sans frontiers. 2013a. "Building a Solidarity City Conference." Accessed February 2, 2016. http://www.solidarityacrossborders.org/en/building-a-solidarity-city-conference-november-23–24.

Cité sans frontiers. 2013b. "Community Statement: 'Ensemble contre la Charte xénophobe'" (Together against the Xenophobic Charter). Accessed December 4, 2015. http://www.solidarityacrossborders.org/en/community-statement-%E2%80%9Censemble-contre-la-charte-xenophobe%E2%80%9D-together-against-the-xenophobic-charter.

Coleman, Nathaniel. 2013. "Utopian Prospect of Henri Lefebvre." *Space and Culture* 16(3): 349–363.

Darling, Jonathan and Vicki Squire. 2013. "Everyday Enactments of Sanctuary: The UK City of Sanctuary Movement." In *Sanctuary Practices in International Perspectives: Migration, Citizenship and Social Movements*, edited by Randy K. Lippert and Sean Rehaag, 191–204. Abingdon: Routledge.

Derrida, Jacques. 2001. *On Cosmopolitanism and Forgiveness*, translated by Mark Dooley, and Michael Hughes. London: Routledge.

Harvey, David. 2005. *A Brief History of Neoliberalism*. Oxford: Oxford University Press.

Harvey, David. 2012. *Rebel Cities: From the Right to the City to the Urban Revolution*. London: Verso.

Hawel, Marcus. 2006. "Negative Kritik und bestimmte Negation: Zur praktischen Seite der kritischen Theorie." In *Aufschrei der Utopie: Möglichkeiten einer anderen Welt*, edited by Marcus Hawel and Gregor Kritidis, 98–116. Hannover: Offizin-Verlag.

Isin, Engin. 2002. *Being Political: Genealogies of Citizenship*. Minneapolis, MN: University of Minnesota Press.

Isin, Engin. 2007. "City State: Critique of Scalar Thought." *Citizenship Studies* 11(2): 211–228.

Isin, Engin. 2008. "Theorizing Act of Citizenship." In *Act of Citizenship*, edited by Engin Isin and Greg Nielsen, 15–43. New York: Zed Books.

Johnson, Heather. 2012. "Moments of Solidarity, Migrant Activism and (Non)Citizens at Global Borders: Political Agency at Tanzanian Refugee Camps, Australian Detention Centres and European Borders." In *Citizenship, Migrant Activism and the Politics of Movement*, edited by Peter Nyers and Kim Rygiel, 109–128. London: Routledge.

Kelz, Rosine. 2015. "Political Theory and Migration: Concepts of Non-Sovereignty and Solidarity." *Movements* 1(2): 1–18. Accessed February 10, 2016. http://movements-journal.org/issues/02.kaempfe/03.kelz–political-theory-migration-non-sovereignty-solidarity.html.

Lefebvre, Henri. 1996. *Writing on Cities*, translated by Eleonore Kofman and Elizabeth Lebas. Oxford: Blackwell.

Linebaugh, Peter. 2008. *The Magna Carta Manifesto: Liberties and Commons for All*. Berkley, CA: University of California Press.

McDonald, Jean. 2012. "Building a Sanctuary City: Municipal Migrant Rights in the City of Toronto." In *Citizenship, Migrant Activism and the Politics of Movement*, edited by Peter Nyers and Kim Rygiel, 129–145. London: Routledge.

Mendieta, Eduardo. 2010. "The City to Come: Critical Urban Theory as Utopian Mapping." *City* 14(4): 442–447.

Merrifield, Andy. 2002. *Metromarxism: A Marxist Tale of the City*. New York: Routledge.

No One Is Illegal. 2012. "#May1TO, May Day: Solidarity City! Status for All! Decolonize Now!" March 1. Accessed February 10, 2016. http://toronto.nooneisillegal.org/MayDay2013.

No One Is Illegal. 2013. Email circular, November 11.

Nyers, Peter. 2010. "No One Is Illegal between City and Nation." *Studies in Social Justice* 4(2): 127–143.

Nyers, Peter and Kim Rygiel. 2012. "Introduction." In *Citizenship, Migrant Activism and the Politics of Movement*, edited by Peter Nyers and Kim Rygiel, 1–19. London: Routledge.

O'odham Newswire. 2010. "1st Nation and Migrants Oppose Sb1070 Demand Dignity, Human Rights, and End to Border Militarization." May 21. Accessed November 30, 2015. http://oodhamsolidarity.blogspot.ca/2010/05/occupation-of-border-patrol-headquaters.html.

Purcell, Mark. 2002. "Excavating Lefebvre: The Right to the City and Its Urban Politics of the Inhabitant." *GeoJournal* 58(2–3): 99–108.

Rancière, Jacques. 1999. *Dis-Agreement: Politics and Philosophy*. Minneapolis, MN: University of Minnesota Press.

Rancière, Jacques. 2004. "Introducing Disagreement." *Journal of the Theoretical Humanities* 9(3): 3–9.

Samers, Michael. 2003. "Immigration and the Spectre of Hobbes: Some Comments for the Quixotic Dr. Bauder." *ACME* 2(2): 210–217.

Sassen, Saskia. 2008. *Territory, Authority, Rights: From Medieval to Global Assemblages*, updated edition. Princeton, NJ: Princeton University Press.

Sassen, Saskia. 2011. *Cities in a World Economy*, 4th edition. Thousand Oaks, CA: Pine Forge.

Sassen, Saskia. 2013. "When the Centre No Longer Holds: Cities as Frontier Zones." *Cities* 34: 67–70.

Scheeres, Julia. 2001. "Borderhack: Barbed and Unwired." *Wired*, 23 August. Accessed November 29, 2015. http://archive.wired.com/culture/lifestyle/news/2001/08/45921.

Sharma, Nandita. 2013. Discussion at the International Workshop "Translating Welfare and Migration Policies in Canada and Germany: Transatlantic and Transnational Perspectives in Social Work," Frankfurt, October 18.

Sharma, Nandita and Cynthia Wright. 2008–9. "Decolonizing Resistance, Challenging Colonial States." *Social Justice* 35(3): 120–138.

Squire, Vicki and Jennifer Bagelman. 2012. "Taking Not Waiting: Space, Temporality and Politics in the City of Sanctuary Movement." In *Citizenship, Migrant Activism and the Politics of Movement*, edited by Peter Nyers and Kim Rygiel, 146–164. London: Routledge.

Standing, Guy. 2011. *The Precariat: The New Dangerous Class*. London: Bloomsbury Academic.

Stierl, Maurice. 2012. "'No One Is Illegal!' Resistance and the Politics of Discomfort." *Globalizations* 9(3): 425–438.

Torpey, John. 2000. *The Invention of the Passport: Surveillance, Citizenship and the State*. Cambridge: Cambridge University Press.

Walia, Harsha. 2013. *Undoing Border Imperialism*. Oakland, CA: A. K. Press.

Walters, William. 2006. "No Border: Games with(out) Frontiers." *Social Justice* 33(1): 21–39.

Wells, H. G. 1959 [1905]. *A Modern Utopia and Other Discussions: The Works of H. G. Wells*, Atlantic edition, Volume IX. London: T. Fischer Unwin.

8 Conclusion

A few years ago, when I presented some of the ideas for this book to my colleague, the political scientist Fabian Georgi, he commented that my critique of borders and migration provides a view into a narrow "crack, a fissure, which can be pried open to show the inhumanity of the whole system and the need to radically change it." He is right. The infringement on individual freedom, the injustice and oppression, and the denial of a future that borders impose cannot be understood in isolation from global capitalism and its colonial history, or the continuing oppression of people based on citizenship, origin, racial markers, gender, or caste ascriptions. The demand for freedom of migration is only one step – albeit an important one – towards human liberation. The preceding chapters offer more than a critique of these narrow topics; they also criticize contemporary global society as a whole.

In his own work, Georgi (2014) puts the calls for open borders and no border on a par with the massive social transformation that followed the abolition of slavery or feudalism. Both of these developments were milestones in history that illustrate how political and social transformation could occur: on the one hand, they required the practical engagement of everyday politics, which enabled advocates to operate within the confines of existing legal and political structures. On the other hand, the abolition of slavery or feudalism demanded the willingness to embark on a path towards an uncertain world. This dual approach to political and social transformation mirrors what I have described as contingent possibility and possibilia, and they must be pursued simultaneously.

Critics may say that activists and scholars contradict themselves when they simultaneously pursue contingent possibilities and possibilia. It is inconsistent, they may argue, to ask the nation state to accommodate migrants by granting them domicile citizenship and simultaneously chip away at the state's legitimacy to grant or deny rights to migrants. Indeed, these positions are contradictory. However, once we let go of the idea that politics – or human life for that matter – must always fit into the straitjacket of consistency and that progress must always be linear, we can accept that contradiction is a necessary moment of social transformation.

This realization has motivated activists to pursue various levels of possibility simultaneously. They see these levels not as mutually exclusive but as complementary. In the words of activist Harsha Walia (2013, 99): "Sustaining a connection between the daily grind of community organizing and broader Left struggles is necessary in order to maintain an expansive political perspective and to stay inspired." This sentiment is echoed by Syed Khalid Hussan (2013, 283), who declares that "we must show that what we bring is both a vision for the future and a way to make things a little better in the present." The contradictions inherent in the different levels of possibility do not paralyze action but propel it. Likewise, critical scholars have long embraced the contradictions between the material world and abstract ideas, and between actual circumstances and what is possible. They must also recognize that the contradiction between contingent possibilities and possibilia is a productive force of societal transformation.

These activists and scholars, however, should also be aware that the dialectical pendulum can swing in an unintended direction. The abolition of slavery and feudalism also produced violence, revolt, and war and was proceeded by a form of capitalism that is also highly problematic. In the same way, the pursuit of open borders and no border will have inadvertent consequences. This pursuit not only gives "hope" (Bloch 1985 [1959]; Harvey 2000) but is also "disconcerting" (Purcell 2002, 100); it can become a dream or a nightmare, and produce mass good as much as mass harm (Hiebert 2002). Since the dialectic of societal transformation is always open, we must stay continuously and critically engaged with immediate practical responses *and* with a larger vision.

Inevitably, humanity will create a world that we cannot fathom from our current viewpoint. In this world, the concepts we cherish will no longer hold true, at least not in the same way. Ironically, this prospect also means that we should not take for granted any of the ideas and concepts that I discussed in the preceding chapters.

One of these ideas relates to the territorial scale at which societies organize themselves. The book started out emphasizing the national scale. I showed how national borders exclude and disenfranchise people, and how border regimes are responsible for the deaths of a staggering number of migrants. Later in the book, I switched to the urban scale to illustrate how migrants can belong to urban communities. However, the nation and the city are not the only scales that define communities to which we can belong. The notion of sanctuary for migrants and refugees, for example, is not only an urban phenomenon, but also exists at other scales and in non-urban contexts (Lippert and Rehaag 2013). While the urban may provide an important "strategic site" (Sassen 2013, 69) to mitigate the illegalization of migrants, the corresponding activist movements are contained neither at national nor urban scales. For example, No One Is Illegal is not only active in large cities but it also fights for Indigenous land claims and Indigenous justice in regions that are not typically considered urban or national (Walia 2013). Even the city itself is not a homogenous territory but consists of particular sites where sanctuary practices are enacted and vast urban spaces where they are absent (Young 2010). Neither the national nor the urban must be seen as natural or the only scales of belonging.

I believe that the question of territory and scale of belonging will be an important matter on the road towards possibilia. Some critical urban theorists no longer distinguish between urban and non-urban contexts. The philosopher Henri Lefebvre began his influential book *The Urban Revolution* with the hypothesis that "society has been completely urbanized" (Lefebvre 2003 [1970], 1). Urban theorist Neil Brenner affirms that "the urban can no longer be viewed as a distinct, relatively bounded site; it has instead become a generalized, planetary condition" (Brenner 2011, 21). And citizenship scholar Engin Isin (2007, 212) suggests that nation states and empires "are kept together by practices organized around and grounded in the city." If the urban encompasses all of society, then it may no longer be useful as an analytical category. The same applies to the national category. Historically, the national scale is a relatively recent phenomenon, and we are observing how this category is being reworked currently in Europe, recently in the former Soviet Union, and through ongoing practices of globalization. In possibilia, both the urban and the national categories may be obsolete.

Another concept that requires rethinking is that of migration. The field of mobility studies has suggested that scholarship should focus on the general movement of persons rather than overemphasizing the particular case of migration across international borders (Cresswell 2006; Urry 2000). The geographer Ben Rogaly (2015, 528) recently suggested that "someone who may once have migrated across international borders does not necessarily see that as the most significant moment in their life," and that moving within the country can be more important to this person. This may be so in cases, for example, when border crossings occur in childhood and subsequent moves within the country are associated with important life experiences and events. However, the main conclusion that I draw in this book is that the focus must remain on migration across international borders, because this type of migration is linked to the exclusion, disenfranchisement, and death of people. Migrant fatalities may look like accidents – the sinking of a boat, a car crash, or losing one's way in the desert – but, clearly, it is not. Rather, the situations leading to these fatalities were created by conscious political decisions and their implementation, and migrants own responses to these decisions.

Critical scholars are well aware that the very concept of the migrant is a construct of international bordering practices in the first place. Being a "migrant" implies that one has crossed an international border (Sharma 2006; Anderson et al. 2009). This realization, however, does not mean that we should no longer study border regimes or criticize border practices.

An interesting example of how migration is currently rethought is found in the notion of "autonomy of migration." This notion suggests that migrants respond to border regimes trying to inhibit their migration in autonomous and creative ways. Autonomous migrants exercise freedom of migration, although this freedom is formally not granted to them. Their autonomy, in turn, forces the state into the defensive. Lacking control over the migrations of populations, the state and other actors must adapt border practices, adjust their mobility management regimes, and develop new technologies to control borders and migration (Casas-Cortes et al. 2015; Nyers 2015; Mudu and Chattopadhyay 2016). In this way, the notion of autonomy of migration describes an inherently dialectical practice. Autonomous migration consists of a dynamic relationship between people who are enacting their freedom of migration and the forces that seek to constrain this freedom. Moreover, autonomous migration is not a mindless response to forces that are out of the migrants' control, but rather a political activity. The autonomous migrants who board the ships, trains, buses, and trucks heading to America, Australia, and Europe act politically in the same way as the people who participated in the 2006 protests in Chicago and the May Day parade in Toronto. By acting politically, these migrants not only alter immediate border politics and practices, but they also hold the key to evoke possibilia.

The concept of freedom, too, needs to be viewed critically. The liberal concept of freedom as the individual autonomy to decide is problematic when it is used to reaffirm unjust economic practices and oppressive political structures (Harvey 2009). Other interpretations of the concept of freedom are also highly context particular. The freedom of migration is a topic of discussion in this book only because this freedom is denied to many people. Yet, most people probably would prefer not to exercise this freedom – they would prefer to stay where they are. They exercise this freedom only because they have no other choice but to migrate to escape war, injustice, poverty, or lack of opportunity. They migrate to stay alive or to improve their lives. The freedom of migration goes hand in hand with the freedom to stay. On the one hand, the freedom of migration and the freedom to stay require different sets of politics and different visions. On the other hand, in possibilia, neither the freedom to migrate nor the freedom to stay will be seen as an issue because these freedoms will be taken for granted.

With *Migration Borders Freedom*, I have tried to bring back a sense of utopian possibility that seems to have gotten lost in the scholarship and politics of recent decades. This possibility includes the freedom of migration and the freedom to belong. The immediate goal, however, must be to stop the murderous practices that caused the death of a two-year-old toddler off the coast of Greece on January 2, 2016; the drowning of 58 people on April 11, 2013, in the Sundra Straight; the abandoning of a pregnant mother in the Arizona desert; and the killing of people whose bodies fill mass graves in the jungles of Thailand and Malaysia. We must not wait for a perfect utopian solution to end this suffering and death, but the inspiration of a more equitable and free possibilia will lead us in the right direction.

References

Anderson, Bridget, Nandita Sharma, and Cynthia Wright. 2009. "Why No Borders?" *Refuge* 26(2): 5–18.

Bloch, Ernst. 1985 [1959]. *Das Prinzip Hoffnung*. Frankfurt/Main: Suhrkamp.

Brenner, Neil. 2011. "What Is Critical Urban Theory?" In *Cities for People, Not for Profit: Critical Urban Theory and the Right to the City*, edited by Neil Brenner, Peter Marcuse, and Margit Mayer, 11–23. Oxon: Routledge.

Casas-Cortes, Maribel, Sebastian Cobarrubias, and John Pickles. 2015. "Riding Routes and Itinerant Borders: Autonomy of Migration and Border Externalization." *Antipode* 47(4): 894–914.

Cresswell, Tim. 2006. *On the Move: Mobility in the Modern Western World*. New York: Routledge.

Georgi, Fabian. 2014. "Was ist linke Migrationspolitik?" In *Luxemburg Gesellschafts-analyse und linke Praxis*. Berlin: Rosa Luxemburg-Stiftung. Accessed January 4, 2016. http://www.zeitschrift-luxemburg.de/was-ist-linke-migrationspolitik.

Harvey, David. 2000. *Spaces of Hope*. Berkeley, CA: University of California Press.

Harvey, David. 2009. *Cosmopolitanism and the Geographies of Freedom*. New York: Columbia University Press.

Hiebert, Daniel. 2002. "A Borderless World: Dream or Nightmare?" *ACME* 2(2): 188–193.

Hussan, Syed Khalid. 2013. "Epilogue." In *Undoing Border Imperialism*, edited by Harsha Walia, 277–281. Oakland CA: A. K. Press.

Isin, Engin. 2007. "City State: Critique of Scalar Thought." *Citizenship Studies* 11(2): 211–228.

Lefebvre, Henri. 2003 [1970]. *The Urban Revolution*, translated by Robert Bononno. Minneapolis, MN: University of Minnesota Press.

Lippert, Randy K. and Sean Rehaag, eds. 2013. *Sanctuary Practices in International Perspectives: Migration, Citizenship and Social Movements*. Abingdon: Routledge.

Mudu, Pierpaolo and Sutapa Chattopadhyay. 2016. *Migrations, Squatting and Radical Autonomy*. London: Routledge.

Nyers, Peter. 2015. "Migrant Citizenship and Autonomous Mobilities." *Migration, Mobility, and Displacement* 1(1): 23–38.

Purcell, Mark. 2002. "Excavating Lefebvre: The Right to the City and Its Urban Politics of the Inhabitant." *GeoJournal* 58(2): 99–108.

Rogaly, Ben. 2015. "Disrupting Migration Stories: Reading Life Histories through the Lens of Mobility and Fixity." *Environment and Planning D: Society and Space* 33(3): 528–544.

Sassen, Saskia. 2013. "When the Centre No Longer Holds: Cities as Frontier Zones." *Cities* 34: 67–70.

Sharma, Nandita. 2006. *Home Economics: Nationalism and the Making of "Migrant Workers" in Canada*. Toronto: University of Toronto Press.

Urry, John. 2000. *Sociology beyond Societies: Mobilities for the Twenty-First Century*. London: Routledge.

Walia, Harsha. 2013. *Undoing Border Imperialism*. Oakland, CA: A. K. Press.

Young, Julie E. E. 2010. "'A New Politics of the City': Locating the Limits of Hospitality and Practicing the City-as-Refuge." *ACME* 10(3): 534–563.

Index